SpringerBriefs in Applied Sciences and Technology

SpringerBriefs present concise summaries of cutting-edge research and practical applications across a wide spectrum of fields. Featuring compact volumes of 50 to 125 pages, the series covers a range of content from professional to academic.

Typical publications can be:

- A timely report of state-of-the art methods
- An introduction to or a manual for the application of mathematical or computer techniques
- A bridge between new research results, as published in journal articles
- A snapshot of a hot or emerging topic
- An in-depth case study
- A presentation of core concepts that students must understand in order to make independent contributions

SpringerBriefs are characterized by fast, global electronic dissemination, standard publishing contracts, standardized manuscript preparation and formatting guidelines, and expedited production schedules.

On the one hand, **SpringerBriefs in Applied Sciences and Technology** are devoted to the publication of fundamentals and applications within the different classical engineering disciplines as well as in interdisciplinary fields that recently emerged between these areas. On the other hand, as the boundary separating fundamental research and applied technology is more and more dissolving, this series is particularly open to trans-disciplinary topics between fundamental science and engineering.

Indexed by EI-Compendex, SCOPUS and Springerlink.

More information about this series at http://www.springer.com/series/8884

Joao Alexandre Lobo Marques ·
Francisco Nauber Bernardo Gois ·
José Xavier-Neto · Simon James Fong

Predictive Models for Decision Support in the COVID-19 Crisis

 Springer

Joao Alexandre Lobo Marques ⓘ
Laboratory of Neuroapplications
University of Saint Joseph
Macau, Macao

José Xavier-Neto ⓘ
Government Intelligence Cell
Secretary of Health of the Government
of the State of Ceara
Fortaleza, Brazil

Francisco Nauber Bernardo Gois ⓘ
Machine Learning Department
Secretary of Health of the Government
of the State of Ceara
Fortaleza, Brazil

Simon James Fong ⓘ
Department of Computer
and Information Science
University of Macau
Macau, Macao

ISSN 2191-530X   ISSN 2191-5318 (electronic)
SpringerBriefs in Applied Sciences and Technology
ISBN 978-3-030-61912-1   ISBN 978-3-030-61913-8 (eBook)
https://doi.org/10.1007/978-3-030-61913-8

This Springer imprint is published by the registered company Springer Nature Switzerland AG
The registered company address is: Gewerbestrasse 11, 6330 Cham, Switzerland

Contents

Chapter 1
Prediction for Decision Support During the COVID-19 Pandemic

1.1 Introduction

At the moment this book is being written, the world is facing the most significant challenge of modern history: the COVID-19 pandemic. The consequences of the outbreak cannot be precisely evaluated yet, since different countries are going through different strategies to cope with the virus pandemic and this is, obviously, generating completely different results. On the other hand, countries where effective strategies could reduce the infection rates a few months ago, are facing second and third waves of infection, indicating that this will be a long-term battle that will probably end not only when a vaccine will be successfully developed, but when the immunization will be effectively in place, with billions of people protected.

The impacts are strongly significant in public health systems, with unprecedented numbers of mortality rates, exponential increasing rates of ICU occupation in short term and limited capacity of response in a large number of healthcare institutions. Additionally, when considering the perspective of the consequences of the pandemic in the global economy, the facts are also frightful and with long-term impacts, especially for developing countries and more specifically for small and medium enterprises. Societies and governments will need a long time and new strategies and actions to recover from the current outbreak minimizing the impact.

The term COVID-19 is an acronym for "Coronavirus Disease 19" and is the name of the disease caused by the virus strain named SARS-CoV-2 (Severe Acute Respiratory Syndrome Coronavirus 2), which belongs to the class of the coronaviruses. It attacks the respiratory system, but besides a common infection in the respiratory tract, several other consequences may affect the patient directly and indirectly. The immune system, for example, trying to combat the virus, in several cases generates such a strong response that affects the patient's life. Consequences of the cardiovascular system and heart functioning are also being widely reported in the medical literature.

When the virus was firstly reported in Wuhan, China and started its exponential spread, several strategies were adopted in parallel, led by general guidelines provided by the WHO (World Health Organization) and specific efforts developed by local

Governments to deal and respond to this challenge, besides the biotechnology efforts to develop medicines and vaccines to fight against the virus action [13].

From the biomedical engineering and computer science perspective, several approaches are being considered, from the development of automatic diagnostic tools based on the application of AI (Artificial Intelligence) software to analyze X-ray and CT-Scans of the patient lungs, to the use of computer-based solution for prediction using epidemiological data based on linear, nonlinear, machine learning, and deep learning techniques.

The main objective of this book is to present different approaches and techniques for epidemiological time-series prediction, which will be able to provide decision support for Governments and Healthcare decision-makers, with the presentation of a sequence of results based on real data from five countries with some similarities and significant differences.

The intent is not to exhaust the topic, but present a discussion covering from conventional compartmental models (SIR and SEIR) to recently developed Artificial Intelligence (AI) solutions. The mathematical level presented for each technique intends to be sufficient for the reader without an advanced technical background, providing references for the ones who want to follow up in detail about each approach presented.

This introductory chapter aims to create a foundation of concepts related to the topic of prediction in health care and epidemiology, which are relevant to understand the application of the techniques presented in each following chapter.

1.2   Explanation and Prediction

At first, it is extremely important to understand the difference of two essential concepts that are often mismatched: explanation, which will be clearly related to the traditional methods presented in explanatory statistics; and prediction, which is the application of specific techniques to predict new information from current data and will be the focus of this Book.

This differentiation is a key step to understand when and why there is the necessity to use prediction techniques and also to clearly identify what is the goal and the type of data being considered.

Inferential statistics have been widely used in explanatory analysis and the focus is usually causal [2]. The application of different regression techniques are common, according to the research scope and type of data. Four of the most used approaches are presented in Sect. 1.2.1.

1.2.1 Explanatory Approaches—Data Regression in Epidemiology

The applications of different regression models are common approaches for dealing with large amounts of data in epidemiology and each of them presents a suitable condition to be applied, depending on the dependent variable and its scale of measurement [2, 7].

The first regression model is known as Linear Regression (LR), which has the simplest mathematical formulation and is considered when analyzing continuous outcomes or explanatory variables, for example, the body temperature measured in degree Celsius or systolic blood pressure, measured in millimeters of Hg. In epidemiology, since the number of variables tends to be large, usually it is common to expand the analysis to multiple linear regression, considering several explanatory variables [11].

Logistic Regression (LogR), on the other hand, is used in epidemiology when dealing with binary outcomes. This means that when the output variable must be a classifier (yes/no), this regression shall be considered. For example, when several explanatory variables are considered to determine if the patient will develop a disease, or even more detailed if the inputs can be classified to indicate the risk of mortality or not [4].

The third approach is the Cox Regression (CoxR), which is considered when time is an essential part of the analysis. These are known as time-to-event data. Normally, clinical trials and cohort studies observe the patients during different windows of time as a result of several research restrictions. When this happens, there is the concept of time-to-event data, also known as survival time data, where the time T until an event occurs is used as the desired output, instead of a binary variable (yes/no), as previously considered for the LogR [5].

Finally, the fourth regression model presented in this introduction is the Poisson regression (PR), which is commonly applied when the output variable is a frequency, a count, or a rate. It can also be considered when using the time-to-event data but, differently from the previous CoxR, the responses are not one single value, but a sequence of it, assuming the occurrence of a Poisson process, considering that the waiting times between successive events are independent and exponentially distributed [6].

As clearly stated, depending on the universe being studied and its corresponding variables, the selection of these explanatory approaches will support the analysis following a causal perspective. For prediction (and specifically, forecasting), other approaches can be considered to enrich the data processing and analysis.

1.3 Prediction for Decision Support

The application of statistical exploratory analysis, forecasting, and prediction techniques are of great value for decision support in several areas of knowledge and, more specifically, applied to healthcare systems management.

Prediction can be considered a general concept related to the estimation of outcomes for unseen data. For that, it is necessary to define a model and train with an existing data x, which will result in an estimator $f(x)$ which will be able to predict outcomes based on new occurrences of data x'.

Forecasting, on the other hand, can be classified as a special case of prediction, based specifically on temporal data, i.e., generating outcomes about the future, based on time-series.

Thus, the only difference between prediction and forecasting is that we consider the temporal dimension. An estimator for forecasting has the form $f(x_1, \ldots, x_t)$ where $x_1, \ldots, x_t$ indicate historic measurements at time points $1, \ldots, t$, while the estimate relates to time point $t + 1$ or some other time in the future. Since the model depends on previous observations, this is generally classified as an autoregressive model.

According to the previous definition, it becomes clear that forecasting requires not only the existence of historical data, but also data enough to contain relevant information about previous events. This brings the light of the discussion to the reliability of the considered data set under analysis, which will be dependent on the impact of technical and nontechnical influences to avoid bias during data collection and, consequently, affect data interpretation.

Although several models of prediction have been developed for a wide number of applications, no prediction can be considered certain as future events normally depend on a large number of factors that will seldom repeat under the same circumstances.

1.3.1 Scope and Time Spam

Two main aspects must be considered by the decision-makers when adopting prediction techniques as a supporting tool. The first one is about the scope of the prediction, while the second aspect is the time spam that might be considered as valid based on the available amount and quality of data.

When talking about the scope of the prediction, it is essential for the decision-maker to critically understand the positive aspects and also the limitations of the available data. This will enforce the scope of any prediction that may be achieved and the reliability of the prediction will be based on that. Considering the COVID-19 pandemic, several possibilities or predictions might be of interest. For example, any prediction directly related to the epidemiological data, such as the number of new infections per day, depends on the logistic control of executing new tests, processing the results, officially release them to the public and update a centralized database. Without a clear process and the proper execution, the considered database might reflect wrong or biased trends and any prediction that will result from the computational system/model will not be able to be validated or, even worse, will support the public healthcare system administrators on making erroneous decisions.

One example of bias that might occur which may affect the scope of the prediction is when we find differences in data collection or even in the process of collecting or processing diagnostic tests. This, not rarely, can be found in certain healthcare systems, mainly located in developing countries. In some healthcare units located in more structured locations (wealthy areas, for example), data are generated faster, while in the poorest regions this process is sometimes neglected, not only because of lack of interest, but sometimes because health teams keep struggling with lack of resources and this may lead them to a less efficient process control. These situations will certainly result in biased databases and reflect in poor or invalid predictions.

For the second aspect related to the time span of the prediction, the major concern is based on the available data, what is the forecast window that should be considered so the prediction will be valid? The discussion here should be connected to the specific infection-recovering window of the disease under analysis. In the case of the COVID-19, available data indicate that persons with mild to moderate symptoms remain infectious no longer than 10 days after symptom onset. Patients with more severe to critical illness will remain infectious no longer than 20 days after symptom onset. Also, recovered patients can continue to shed detectable SARS-CoV-2 RNA in upper respiratory specimens for up to 3 months after illness onset [1]. In this case, a starting point to set an admissible time spam for the COVID-19 pandemic would be to consider a forecast window between 5 and 20 days, so it would be possible to measure what would be the infection time-series behavior on time. Obviously, for these aspects, it is also important to notice that the challenge presented before of different structures of data collection on the same healthcare system will also negatively impact and must be carefully considered as well.

1.4 The Proposed Approach for the Methodology and Results Presentation

Writing a book about prediction techniques for decision support considering a dynamic subject such as the COVID-19 pandemic is a significant challenge, especially because the pandemic is still ongoing while the book is being elaborated, several variables are influencing the COVID-19 pandemic behavior and several publications are being published about the topic worldwide. Nevertheless, the authors consider the book an important contribution, providing a clear methodological approach and a coherent sequence of techniques that will clarify to the reader about the area and make it possible to adopt any of the proposed techniques or go further from them and adapt or develop new prediction models.

To allow a better comprehension and analytical comparison between the approaches presented in this book, Chaps. 2–5 will follow a similar methodology, considering the same data sources and the same group of countries: China, United States, Brazil, Italy, and Singapore. In Chap. 6, a case study about the geographical prediction from the City of Fortaleza, in Brazil, is presented.

Analyzing the behavior of the pandemic, such as the infection rates or new confirmed cases per day, from different countries (or even regions, if the reader, for example, accepts to consider Italy as a representation of Europe, in general), it is clear the presence of different trends. Understanding the temporal behavior of these epidemic curves presented in this introductory section will provide the basic support to understand the positive and negative aspects of each prediction tool for each country in the following chapters.

This book provides a practical approach and discussion, following a logical sequence of different prediction and forecast techniques, starting with the analysis of three compartmental modeling tools in epidemiology (SIR; SEIR and SEIR with Intervention); followed by the application time-series autoregressive approach, ARIMA; Kalman filtering applied for COVID-19 prediction; AI application based on LSTM recurrent neural networks and a solution based on an auto selector of machine learning techniques; and, finally, a case study based on geographic predictor developed for the Government of the State of Ceara, Brazil.

The proposed sequence is presented in Fig. 1.1 and further details about the materials and methods in common for each chapter are discussed in Sect. 1.5.

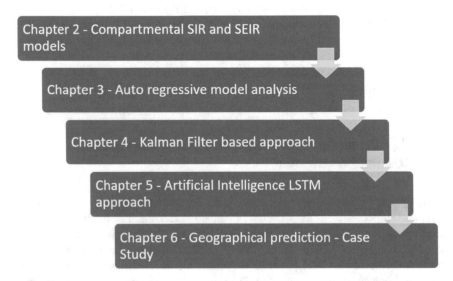

Fig. 1.1 Structure of the book, following a sequence of different approaches for the COVID-19 prediction

Although it was designed a common methodology for different chapters, each of them will be independent and can be considered individually for reading and referencing, according to the proposed technical approach.

1.5   Methodology and Scope Definition for the Data Analysis

In this section, general aspects of the methodology adopted for the analysis throughout the book are presented.

1.5.1   Data Source and Data Selection

First of all, it is important to remind that there are several public databases providing access to data related to the COVID-19 pandemic. The data source considered for this Book was the one provided by the Johns Hopkins University (JHU) and Center for Disease Control (CDC), which are supported also by the World Health Organization (WHO) [13].

After defining the reliable source of our data, five countries were selected for the analysis (from Chap. 2–5): China, United States of America (USA), Brazil, Italy, and Singapore, each of them with specific characteristics.

The epidemic curves of the number of new cases per day for each country from the 1st of February to the 31st of July of 2020 are then presented. The reader must notice that the plots are in different scales of daily number of infections (y-axis) since the numbers in Brazil and the USA are quite larger than the other ones. Italy presented a significant peak in March, Singapore presented a lower peak in April and China presented the first peak of all countries but kept lower numbers of infections because of extreme contention measures of lockdown, traveling restrictions, and quarantine, obtaining a decrease in the epidemic curve in a fast rate, when considering the number of daily new cases.

It is actually very relevant and challenging for computational models to have different epidemic curves and behavior for different countries because this demonstrates the high influence of human decisions interacting with the local epidemic and this will be considered during the model analysis in the following chapters.

China was the first country to detect and report human infection by the SARS-COV-2, in the city of Wuhan. With a wide interconnected economy, the country acted in a fast and responsible way to contain the virus spread, with an impressive response on isolation and management of medical resources for the infected area and this can be clearly verified in Fig. 1.2.

Two other countries, the United States and Brazil, followed a different decision-making approach from their central governments, with the first position of denial of the pandemic and focus more on the economic threats than on the healthcare strategies for the virus spread contention. The situation in both countries is still

Daily New Cases in China

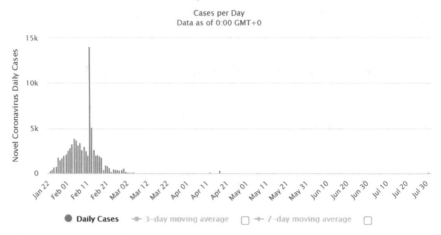

Fig. 1.2 Epidemic curve for China until the end of July, 2020

Daily New Cases in the United States

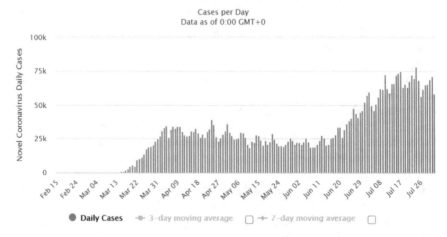

Fig. 1.3 Epidemic curve for the USA until the end of July, 2020

facing a stable number of daily new cases, as can be seen in their epidemic curves presented in Figs. 1.3 and 1.4, respectively.

Italy was selected to represent a typical epidemic curve behavior from Europe (similar to Belgium and Spain, for example) with a severe peak in the number of cases and subsequent strong measures to contain the virus spread, as can be seen in

Daily New Cases in Brazil

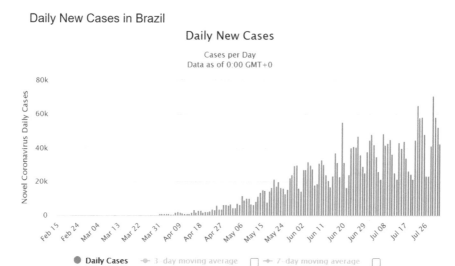

Fig. 1.4 Epidemic curve for Brazil until the end of July, 2020

Daily New Cases in Italy

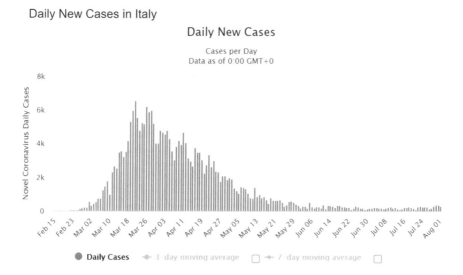

Fig. 1.5 Epidemic curve for Italy until the end of July, 2020

Fig. 1.5. It is important to notice that several measures for containing the virus were released and the risk of new waves is under strong observation.

Finally, Singapore could keep the numbers on a smaller scale than the other countries under discussion, with a peak of less than 1,500 daily new cases in April. Although the numbers are low, it is interesting to check that the risk of new waves

Daily New Cases in Singapore

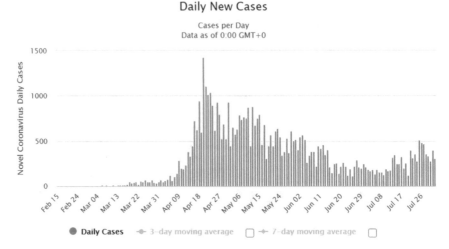

Fig. 1.6 Epidemic curve for Singapore until the end of July, 2020

was always present and it is possible to verify visually a slight increase in the average number of daily new cases in July. The plot is presented in Fig. 1.6.

Despite the fact that different countries followed different evolution of the pandemic spread according to their countermeasure actions, it is essential to understand that, since the COVID-19 pandemic is something new and depends on a large number of variables, the decision-making processes for short and mid-term, extremely needs the support of predictive models as an auxiliary method to support healthcare systems strategies and adapt fast to changes in a short time frame.

1.5.2   Methodology for Validation and Evaluation

1.5.2.1   Cross-Validation

Cross-validation can be considered as a sample selection or division procedure usually adopted when using prediction or machine learning models when the number of samples is limited [3].

It consists of selecting one parameter k that will represent the number of groups that the available data sample shall be split into. The term "k-fold cross-validation" is used to represent that a chosen k will be set for the validation, for example, if $k = 10$, it is called 10-fold cross-validation.

Cross-validation is commonly used in machine learning approaches to estimate the skill of the selected model, such as KNN, Random Forest, or XGBoost. The consequence of using samples with limited size in order to estimate how the model is expected to perform in general when used to make predictions on data not used during the training of the model. Presenting a less biased and less optimistic approach, despite its simplicity, the method provides a good strategy besides the usual division of existing samples into training and testing subsets.

Nevertheless, the decision of using a specific k can generate bias and influence the estimation of the prediction error, which is the key process to evaluate the performance of one model [3]. The usual discussion of selecting a 5-fold or 10-fold approach must be carefully analyzed and results should be compared for validation.

The analysis presented in Chaps. 3–5 will consider a 10-fold cross-validation strategy. This means that for each country 5 time-series subsets will be considered for testing the model and the evaluation criteria are calculated.

1.5.2.2 Accuracy Evaluation Criteria

One of the most important parts of our methodology is to define the criteria for the accuracy evaluation and classification and determine the performance of each proposed model and, consequently, the quality of their respective predictions. Several metrics are presented and discussed in the literature [9] and introductory and detailed discussion can be found in [12, 14].

In this book, for Chaps. 3–5, the authors considered three commonly used metrics in the literature for model accuracy evaluation: R2 Score; Mean Square Error (MSE), and Mean Absolute Error (MAE).

The R-squared score, commonly known as R^2 (and sometimes R2 for the sake of simplified writing), is a statistical measure that shows the dependent variable's variance percentage that collectively determines the independent variable. It is a linear model to measure the relationship strength between the dependent variable and the selected regression model following a percentage scale. Graphically thinking, this metric will find how scattered are the data points around the regression line which is commonly referred to as the coefficient of determination.

Lower R^2 scores indicate that the dependent variable has no variability around its mean, while higher scores imply that the response variable has all the variability around its mean. The higher the score, the better for the model evaluation [10]. It can be expressed as presented in Eq. 1.1:

$$R^2 = \frac{V_M}{V_T},\tag{1.1}$$

where V_M is the variance obtained by the model and V_T is the total variance of the data. This formulation is normally considered on the analysis and prediction of economics and financial time series.

Another approach to estimate the criterion is called Adjusted R^2 Score and is presented in Eq. 1.2.

$$R^2 = 1 - \frac{S_R}{S_T}, \tag{1.2}$$

where S_R is the sum of squares of residuals and S_T is the total sum of residuals, which is proportional to the variance of the data. This formulation is usually considered to evaluate Machine Learning prediction algorithms.

It is essential to notice that, in case of considering this second formulation, the R^2 Score calculation can result in negative values, in the case where S_R is greater than S_T. Also, under some specific conditions, the Score is undetermined.

The Mean Absolute Error (MAE) is the average magnitude of the errors in the set of model predictions. It is presented in Eq. 1.3, considering y_j as the data and $\hat{y}_j$ as the prediction result [15].

$$MAE = \frac{1}{n} \sum_{j=1}^{n} |y_j - \hat{y}_j| \tag{1.3}$$

This is an average on the comparison or difference between the model predictions and actual data, where all individual samples have equal weight. Its value ranges from 0 to infinity and the lower the MAE score will result in a better fit for the proposed model.

Finally, the Mean Square Error (MSE) is presented in Eq. 1.4, considering y_i as the data and $\hat{y}_i$ as the prediction result [15]. It is the most classic method for error calculation and is based on squaring the distance of each data point from the regression line. The squared difference is used to avoid negative values and give more emphasis to larger errors. The smaller MSE possible will give the closer the model is to find its best prediction [8].

$$MSE = \frac{1}{n} \sum_{i=1}^{n} (y_i - \hat{y}_i)^2 \tag{1.4}$$

The discussion among different 5-fold test data results for one country and comparisons about the performance among different countries is performed based on each of these criteria.

1.6 Conclusion

In this chapter, a general presentation about the challenge of adopting prediction techniques to support the decision-making process during the COVID-19 pandemic is presented. A brief discussion about explanation and prediction is provided together with a clear definition of the term forecasting. The methodology proposed by the authors of the book is detailed and the performance criteria for the evaluation of the results are also presented.

In the following chapter, three different compartmental models, which are widely adopted prediction models in epidemiology, are considered for the analysis of the 5 selected countries.

References

1. M.M. Arons et al., Presymptomatic SARS-CoV-2 infections and transmission in a skilled nursing facility, in *New England Journal of Medicine* 382.22(2020), pp. 2081–2090. issn: 15334406. https://doi.org/10.1056/NEJMoa2008457
2. R. Bender, Introduction to the use of regression models in epidemiology, in *Methods in Molecular Biology*, vol. 471 (2009), pp. 179–195. issn: 10643745. https://doi.org/10.1007/978-1-59745-416-2_9
3. T. Fushiki, Estimation of prediction error by usingK-fold crossvalidation, in *Statistics and Computing* 21.2 (2011), pp. 137–146. issn: 09603174. https://doi.org/10.1007/s11222-009-9153-8. https://link.springer.com/article/10.1007/s11222-009-9153-8
4. F.E. Harrell, Case Study in Binary Logistic Regression, Model Selection and Approximation: Predicting Cause of Death (2015), pp. 275–289. https://doi.org/10.1007/978-3-319-19425-7_11
5. F.E. Harrell, Case Study in Cox Regression (2015), pp. 521–533. https://doi.org/10.1007/978-3-319-19425-7_21
6. F.E. Harrell, Introduction to Survival Analysis (2015), pp. 399–422. https://doi.org/10.1007/978-3-319-19425-7_17
7. G. James et al., Statistical Learning (2013), pp. 15-57. https://doi.org/10.1007/978-1-4614-7138-7_2
8. R. Kaundal, A.A. Kapoor, G.P.S. Raghava, Machine learning techniques in disease forecasting: a case study on rice blast prediction, in *BMC Bioinformatics* 7.1 (2006), p. 485. issn: 14712105. https://doi.org/10.1186/1471-2105-7-485. http://bmcbioinformatics.biomedcentral.com/articles/10.1186/1471-2105-7-485
9. J. Li, Assessing the accuracy of predictive models for numerical data: Not r nor r2, why not? Then what?, in *PLOS ONE*, vol. 12.8, ed. by Q. Zhang (2017), p. e0183250. issn: 1932-6203. https://doi.org/10.1371/journal.pone.0183250.
10. J. Lupón et al., Biomarker-assist score for reverse remodeling prediction in heart failure: The ST2-R2 score, in *International Journal of Cardiology* 184.1 (2015), pp. 337–343. issn: 18741754. https://doi.org/10.1016/j.ijcard.2015.02.019. https://pubmed.ncbi.nlm.nih.gov/25734941/
11. D.E. Matthews, Linear regression, Simple, in *Wiley StatsRef: Statistics Reference Online* (Wiley, Chichester, UK, 2014). https://doi.org/10.1002/9781118445112.stat05758. http://doi.wiley.com/10.1002/9781118445112.stat05758
12. M. Schemper, Predictive accuracy and explained variation, in *Statistics in Medicine* 22 (2003)
13. WHO,World Health Organization - (2020). Coronavirus Disease (COVID-19) Situation Reports. https://www.who.int/emergencies/diseases/novel-coronavirus-2019/situation-reports/ (visited on 08/02/2020)
14. C.J. Willmott, On the validation of models, in *Physical Geography* 2 (1981)
15. C.J. Willmott, K. Matsuura, (n.d.) Advantages of the mean absolute error (MAE) over the root mean square error (RMSE) in assessing average model performance (1981). https://doi.org/10.2307/24869236. https://www.jstor.org/stable/24869236

Chapter 2
Epidemiology Compartmental Models—SIR, SEIR, and SEIR with Intervention

2.1 Introduction

The interpretation, analysis, and decision-making processes for dealing with infectious diseases are currently firmly based on mathematical modeling, since physicians and academics started researching about epidemiological models [8]. During the past decades and centuries, at least, several techniques and approaches have been proposed with a tendency of achieving good results depending on the model parameters definition, according to specific characteristics of the diseases. As a classic example, the research performed by Daniel Bernoulli, in 1760, about modeling the smallpox epidemic behavior became a reference. The discussion about inoculation and the risk of deadly infection was still a controversy, but he focused on showing that inoculation was advantageous if the associated risk of dying was less than 11%, which means up to three years, at that time [1]. During the following decades, models and projections have become popular and pervasive as a supporting tool for governance.

Several fields of study already consider mathematical models, many of them adapted to computerized solutions, to support their decision process, using predictions in areas such as weather forecasting, financial behavior, and public health management. This last one is the focus of this book and the application of several models was strongly adopted during recent epidemics of SARS and MERS and, more recently during the global COVID-19 pandemic.

The management control obtained through the adoption of prediction models is not only a matter of perception but a result of the system requirements and the proper model selection, according to these requirements. This is not based on assumptions but on a clear methodology that the decision-maker must follow. Rhodes et al. [8] presented interesting definitions about the concept of prediction, starting with a body of work in social science points to knowing futures as a means to taming the uncertainties of the present, including through an imagined universal "trust in numbers" and finally states that projections close down unknowns into a governable present.

© The Author(s), under exclusive license to Springer Nature Switzerland AG 2021
J. A. L. Marques et al., *Predictive Models for Decision Support in the COVID-19 Crisis*, SpringerBriefs in Applied Sciences and Technology,
https://doi.org/10.1007/978-3-030-61913-8_2

Some essential assumptions or classifications about the system under study are necessary for a correct selection and application of the model. The first essential assumption is if the monitored system is deterministic or stochastic. The first one is also called the compartmental model and assumes that individuals in the population are assigned to different subgroups or compartments, each representing a specific stage of the epidemic. For the stochastic system, it assumes that the epidemiological data follows a probability distribution of potential outcomes by allowing for random variation in one or more inputs over time. Stochastic models depend on the chance variations in the risk of exposure, disease, and other illness dynamics [3].

The transition rates from one class to another are mathematically expressed as derivatives, hence the model is formulated using differential equations. While building such models, it must be assumed that the population size in a compartment is differentiable with respect to time and that the epidemic process is deterministic. In other words, the changes in the population of a compartment can be calculated using only the history that was used to develop the model [3].

One deterministic model widely considered in epidemiology it the Susceptible–Infectious–Removed (SIR) model, which is based on the classification of the individuals into three stages of infection and was introduced almost one hundred years ago by W.O.K. McKendrick and A.G. McKendrick [7]. More details about the model are presented in the following sections. At this point, it is essential to understand that goal is to explain the process of a virus spread in a human community based on vectors; i.e., susceptible individuals who get infected from contagious vectors [11] Beretta1995 schenzle1984age.

A variation and expansion of the SIR model is the Susceptible–Exposed-Infectious–Removed (SEIR) model. This one considers the classification of the individuals into four stages of infection. The introduction of the "Exposed" stage creates a new class for the individuals who may have been in touch with the virus, but cannot be classified as infected yet.

Finally, a modified SEIR model will consider an intervention factor to make it possible to reflect external measures into the infection rates. These models must be considered as a tool to support the decision-making process, but may not reflect the effective rates of infection in medium and long-term analysis. Real-life models are subject to many other external and internal parameters. In the following sections, the equations and parameters of SIR, SEIR, and SEIR with intervention models are presented.

The main objective of this chapter is to present and discuss a series of predictions about the behavior of the pandemic in five countries: China, United States, Brazil, Italy, and Singapore based on their own historical data, considering three compartmental models: SIR, SEIR, and SEIR with Intervention. Positive aspects and limitations of each model are presented, allowing the reader to analyze their practical application.

2.1.1 Literature Review

After an introduction and justification about the adoption of compartmental epidemiological models, a brief literature review is presented focused on recent publications about the application of compartmental models and several possible variations on the analysis and prediction of the COVID-19 pandemic in different countries.

Since the early phases of the COVID-19 outbreak, even before the WHO's declaration of the classification as a pandemic, several works in the literature started to apply different models to try to predict the epidemic behavior of this disease, basically because of two main aspects: the high speed of contamination, as a result of transmission through the air, and the fatality rate, which was oscillating around 4–5% in China.

Wu et al. [10] analyzed data from December 31, 2019 to January 28, 2020, on the number of cases exported from Wuhan internationally (known days of symptom onset from December 25, 2019 to January 19, 2020) to infer the number of infections in Wuhan from December 1, 2019 to January 25, 2020. The authors focused the prediction on the number of cases exported domestically from Wuhan to other major cities in China, such as Beijing, Shanghai, Guangzhou, and Shenzhen. The SEIR model was considered and the basic reproductive number was estimated using Markov Chain Monte Carlo methods. These cities were affected by domestic infections, but the Government measures to close the borders of Wuhan and restrict travels inside China could contain effectively the rates and the contamination rates didn't follow exponential rates as predicted for these major cities.

In Italy, one of the most affected countries in Europe, a new model was proposed to predict the course of the epidemic and analyze influential factors to control its spread [5]. Instead of the traditional models with three, four of five stages of infection, this one proposed eight: susceptible (S), infected (I), diagnosed (D), ailing (A), recognized (R), threatened (T), healed (H), and extinct (E), collectively termed SIDARTHE. The results allow the conclusion that social-distancing measures, which are considered the only measure in several developing countries, need to be combined with effective testing strategies and contact tracing to be able to combat the spread of the COVID-19 disease.

Still considering the outbreak in Italy, Gatto et al. [4] presented a model based on the Susceptible–Exposed–Infected–Recovered (SEIR), considering 107 provinces and the contribution of presymptomatic and asymptomatic transmission, which is relevant specifically for the COVID-19, comparing to other epidemics. The data considered the infections in Italy from February to March 2020, when were implemented progressive restrictions. Adopting an infection rate $R_0 = 3.60$ and adding uncertainty parameters to the model, the results suggest that the sequence of restrictions posed to mobility and human-to-human interactions have reduced transmission by 45% and created a positive impact on the number of hospitalizations and a significant reduction on the number of necessary resources to manage the number of infections.

Another innovative approach was proposed by Samui et al. [9], with a four-compartment model called SAIU (Susceptible-Asymptomatic-Infectious-Unreported Infectious) which is a variation of the traditional SEIR model, with a fine-tuning process to determine some of the parameters, such as the transmission rate. The data analysis was performed until the end of April 2020 and the prediction estimated a peak in India in 60 days, followed by a plateau and no predictions about the outbreak reduction. Actually, the epidemic curves in India are still on a growing trend, even after 90 days of the prediction. This reinforces that strong social contention measures are the only effective action to stop the virus spread and this didn't happen in India at the time the prediction was considered. A plateau of new daily infections is almost impossible to be achieved.

An interesting approach was conducted by Ellison [3], who performed a survey about the application of the classical SIR model and some of its variations. It shows the critical role played by the parameters selection step on these models, such as the heterogeneity in contact rates. A bad parameter selection may lead to bias and poor prediction results. Additionally, important conclusions are completely erroneously estimated such as the interval for obtaining herd immunity and also the quantification of the impact of measures such as social distancing.

Another research considered several approaches of prediction regressions for a group of 8 countries from January to March 2020 [11], when these countries were the most significant outbreaks. Comparing their short-term prediction results and peak predictions, the differences among the models were relatively large. As expected, the logistic growth model estimated a smaller epidemic size than the basic SEIR model. With the addition of new parameters to reflect public health interventions and control measures, the adjusted SEIR model results demonstrated a considerably rapid deceleration of epidemic development. The results demonstrate that contact rate, quarantine scale, and the initial quarantine time and length are important factors in controlling epidemic size and length.

After this literature review considering different approaches on the topic, it is necessary to present the methodology designed for this chapter.

2.2 Materials and Methods

2.2.1 SIR Model

The SIR model is among the most fundamental compartmental representations, and several models are extended of this basic one, including the SEIR case. The SEIR model defines three partitions: S for the amount of susceptible, I for the number of infectious, and R for the number of recuperated or death (or immune) people Stone2000.

The differential equations that describe the SIR model are described in Eqs. 2.1, 2.2, and 2.3, all related to a unit of time, usually in days. Then, at each iteration or

Fig. 2.1 SIR model and the transitions between the compartments

instant of time t, the values of each compartment can be changed (Beretta1995) and (Stone2000).

$$\frac{dS}{dt} = -\frac{\beta I S}{N},$$ (2.1)

$$\frac{dI}{dt} = \frac{\beta I S}{N} - \gamma I,$$ (2.2)

$$\frac{dR}{dt} = \gamma I.$$ (2.3)

The modeling assumes that $S(t) + I(t) + R(t) = N$, where N represents the total population. Then in each iteration t, individuals move from S to I and from I to R.

Equation 2.1 describes the dynamics of the reduction of susceptible individuals, where β is the average number of people who come into contact with another person multiplied by the likelihood of infection in that contact.

Equation 2.2 represents the variation for the I compartment, where the new infected ones according to the rate are added and those who were recovered or died are removed, proportional to the parameter $\gamma = 1/D$, where D is the number of days that one individual stays infected.

The last Eq. 2.8 explains the variation on the compartment of the recovered/ mortality patients, which is also directly proportional to γ.

Figure 2.1 illustrate all compartment transitions, showing the transition rate for each time in the arrows.

This model requires as input the amount of the susceptible, infected, and cured or dead population, all referring to a reference time, called time 0. The parameters are necessary to establish the rates and the model dynamics.

2.2.2 SEIR Model

Because the SIS and SIR model exclusively supports the cases without an incubation period, which is not the case for several classes of contagious infections, Cooke proposed a spread model for the case that after a specific period, the susceptible person can get infectious. This model is named as the SEIR model [2].

The SEIR model differs from the SIR in one compartment, the E representing Exposure, which refers to diseases that are not manifested at the exact moment

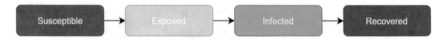

Fig. 2.2 SEIR model with the transitions between the compartments Cooke1979

of infection, having an incubation period. Like COVID-19, which has an ordinary incubation period of 14 days.

The model is defined with four differential equations, described in Eqs. 2.4, 2.5, 2.6, and 2.8. Some small changes are made, starting with the addition of the new Eq. 2.5, which represents the calculation of individuals exposed to the virus [2].

The model added a new parameter, the incubation rate, σ, which is the rate of latent individuals becoming infectious (typical period of incubation is $1/\sigma$).

$$\frac{dS}{dt} = -\frac{\beta I S}{N}, \tag{2.4}$$

$$\frac{dE}{dt} = \frac{\beta I S}{N} - \sigma E, \tag{2.5}$$

$$\frac{dI}{dt} = \sigma E - \gamma I, \tag{2.6}$$

$$\frac{dR}{dt} = \gamma I. \tag{2.7}$$

Analogous to the SIR representation, the sum of the compartments, which are now $S(t) + E(t) + I(t) + R(t) = N$, results in the total population (Fig. 2.2).

The parameters considered for the SEIR model are described below:

- beta (β): probability susceptible–infected contact results in a new exposure;
- gamma (γ): probability of one infected subject gets recovered;
- sigma (Σ) or alpha (α): probability of one exposed person becomes infected.

For the analysis, it is also considered a $R_0 = 3.954$ as a standard rate of reproduction.

2.2.3 SEIR with Intervention Model

Several adaptations of the conventional SIR and SEIR models are vastly presented in the literature, some including new compartments while others focusing on the adjustment of the equations and corresponding parameters.

For this model, the rate of contamination reproduction is a function of time, based on a parameter L and a constant k, as presented in Equation

$$\frac{dR}{dt} = \frac{R_0}{(1 + (\frac{t}{L})^k}. \tag{2.8}$$

Initially is considered $R_0 = 3.954$ as a standard rate of reproduction. Increasing the parameter L will reduce the denominator not influence substantially the variation of the rate of reproduction. On the opposite, decreasing L will increase the denominator and reduce the rate of the virus reproduction. This can represent the application of contention measures such as quarantine and lockdown.

2.3 Results and Discussion

For the results presentation and discussion, the three compartmental models are considered for the prediction results in the following sequence: SIR, SEIR, and SEIR with Intervention.

Despite the fact that SEIR is an extension of the SIR method, the results are presented in two independent subsections, so the reader can have a clear perception about the positive and negative aspects of each model, for the five selected countries proposed in the methodology.

Additionally, since the usual models are based on a group of static parameters for the differential equations, the results of the modified SEIR model are also presented with the variation of the intervention rate L, that will allow us to change the rate of infection R_0, reflecting external changes, such as social isolation measures, in the graphics and, as a consequence, in the analysis.

2.3.1 SIR Model—Results and Discussion

The results for the SIR model are presented following the sequence established for the country analysis in the whole book. The first country is China and the results are presented in Fig. 2.3. The second analysis is performed for the United States and the SIR plots are presented in Fig. 2.4. In Fig. 2.5 are the plots with the results of the SIR model based on the Brazilian dataset. The fourth country is Italy (Fig. 2.6) followed by Singapore (Fig. 2.7), which is the last analysis.

With the aim to understand the impact of parameter selection for the SIR compartmental model, a total of nine plots are presented for each country considering a variation of the γ and β parameters. First, a fixed $\gamma = 0.2$, which corresponds to an infection duration of 5 days, is selected and the three different values for β are considered: $\beta = [1, 2, 3]$. This means that the proposed simulation will consider different possibilities for the capacity of the virus to spread through personal contacts and its inherent rate of transmission. The same process is then repeated for $\gamma = 0.1$, corresponding to an infection duration of 10 days and $\gamma = 0.066$, which corresponds to an infection duration of 15 days, which is also used as a reference.

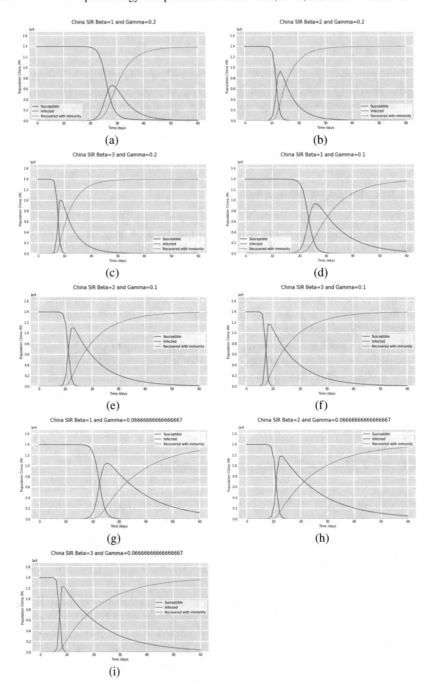

Fig. 2.3 Results for China: Prediction based on the SIR model considering the variation of the parameters γ and β. Figures **a, b**, and **c** $\gamma = 0.2$ and $\beta = [1, 2, 3]$, respectively. Figures **d, e**, and **f** $\gamma = 0.1$ and $\beta = [1, 2, 3]$, respectively. Figures **g, h**, and **i** $\gamma = 0.066$ and $\beta = [1, 2, 3]$, respectively

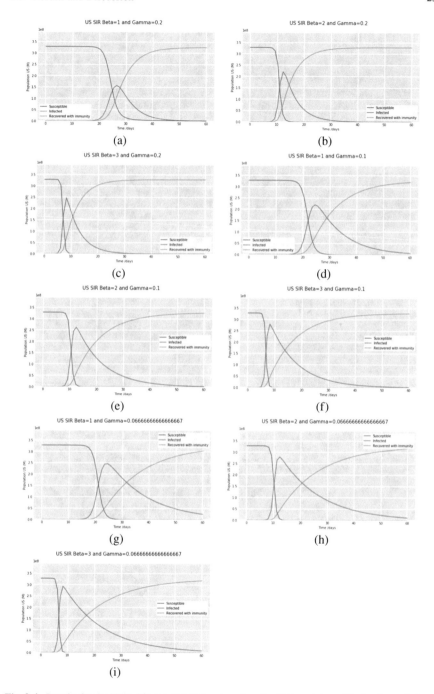

Fig. 2.4 Results for the United States: Prediction based on the SIR model considering the variation of the parameters γ and β. Figures **a**, **b**, and **c** $\gamma = 0.2$ and $\beta = [1, 2, 3]$, respectively. Figures **d**, **e**, and **f** $\gamma = 0.1$ and $\beta = [1, 2, 3]$, respectively. Figures **g**, **h**, and **i** $\gamma = 0.066$ and $\beta = [1, 2, 3]$, respectively

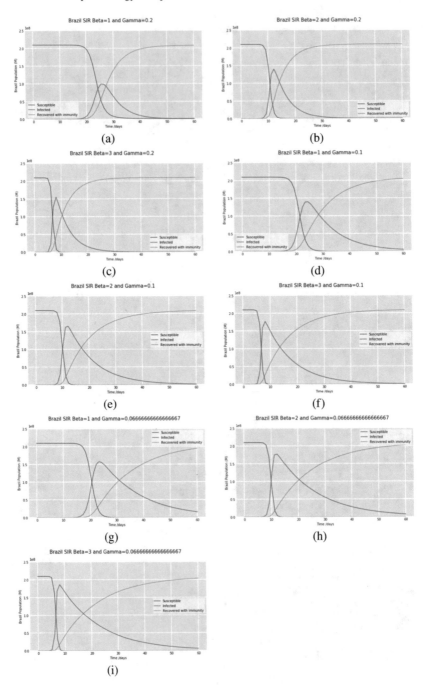

Fig. 2.5 Results for Brazil: Prediction based on the SIR model considering the variation of the parameters γ and β. Figures **a**, **b**, and **c** $\gamma = 0.2$ and $\beta = [1, 2, 3]$, respectively. Figures **d**, **e**, and **f** $\gamma = 0.1$ and $\beta = [1, 2, 3]$, respectively. Figures **g**, **h**, and **i** $\gamma = 0.066$ and $\beta = [1, 2, 3]$, respectively

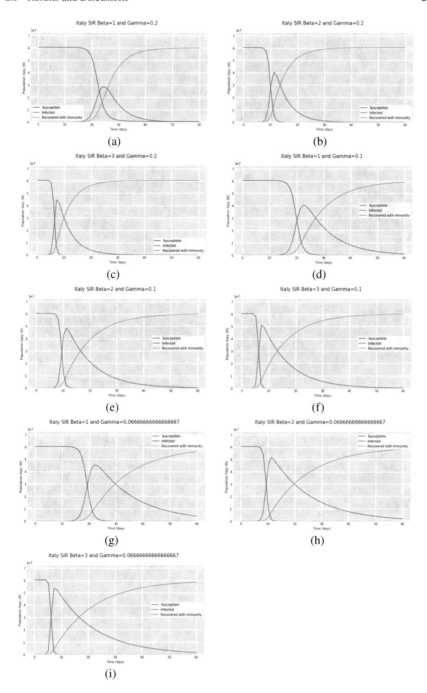

Fig. 2.6 Results for Italy: Prediction based on the SIR model considering the variation of the parameters γ and β. Figures **a**, **b**, and **c** $\gamma = 0.2$ and $\beta = [1, 2, 3]$, respectively. Figures **d**, **e**, and **f** $\gamma = 0.1$ and $\beta = [1, 2, 3]$, respectively. Figures **g**, **h**, and **i** $\gamma = 0.066$ and $\beta = [1, 2, 3]$, respectively

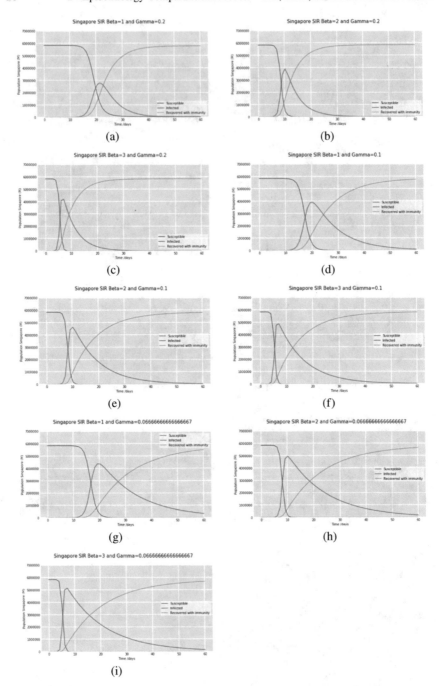

Fig. 2.7 Results for Singapore: Prediction based on the SIR model considering the variation of the parameters γ and β. Figures **a**, **b**, and **c** $\gamma = 0.2$ and $\beta = [1, 2, 3]$, respectively. Figures **d**, **e**, and **f** $\gamma = 0.1$ and $\beta = [1, 2, 3]$, respectively. Figures **g**, **h**, and **i** $\gamma = 0.066$ and $\beta = [1, 2, 3]$, respectively

2.3.2 SEIR Model—Results and Discussion

Similarly to what was presented in the previous section, the results for the SIR model follow the same sequence of countries: China (Fig. 2.9), the United States (Fig. 2.14), Brazil (Fig. 2.15), Italy (Fig. 2.16), and Singapore (Fig. 2.17) (Figs. 2.8, 2.10, 2.11, 2.12, 2.13).

With the aim to understand the impact of parameter selection for the SEIR compartmental model, a total of three plots are presented for each country considering a variation of the γ and β parameters. First, a fixed $\gamma = 0.2$, which corresponds to an infection duration of 5 days is selected and three different values for the infection rate are considered: $\beta = [2.5, 3.0, 3.9]$. This means that the proposed simulation will consider different possibilities for the capacity of the virus to spread through personal contacts and its inherent rate of transmission.

In order to understand the SEIR epidemiological modeling and respective curves, it's important to identify key parts of the graphical data. The most relevant information to be extracted from the plots and considered in the interpretation are listed below:

- the fraction of population (the percentage) of the peak of the infection curve;
- the number of days to achieve the peak of the infection curve;
- the fraction of population (the percentage) of the peak of the exposed curve;
- the number of days to achieve the peak of the exposed curve;
- the fraction of population (the percentage) when the susceptible curve crosses with the recovered curve;
- the number of days to the susceptible curve crosses the recovered curve;
- the fraction of population (the percentage) when the susceptible curve stabilizes;
- the fraction of the population (the percentage) when the recovered curve stabilizes.

The following Figs. 2.14, 2.15, 2.16, and 2.17 are a sequence of results for the five countries considered in this book. Specifically in this section, we are going to exchange the first and second countries from the original sequence and the first group of plots will be from the United States, which is the target of our example here, followed by China. After that, the original sequence is preserved with Brazil, Italy, and Singapore as the last one.

The SEIR modeling for the United Stated is presented in Fig. 2.14. The interpretation here will be based on visual analysis and considering the perception of a public healthcare manager. Analyzing the example, the first interpretation, as expected, is that increasing the R_0 will impact the speed of the infection and reduce the number of days to achieve the peak of infected and exposed. Besides the fraction of the population peaks for the infected and exposed curves also increased according to the rate of infection. In all three cases, the susceptible curve converged to a very low fraction of the population while the recovered/deceased curve converged to almost the total of the population.

The same approach can be followed for the analysis of each country and since the SEIR parameters are the same, the interpretation will be quite similar. It is important to notice that selecting the proper R_0 should be influenced not only by the epidemic

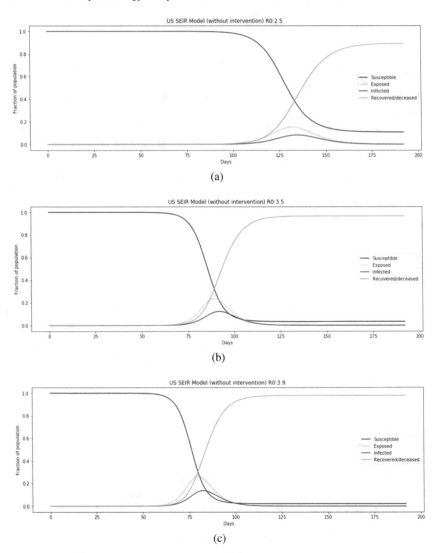

Fig. 2.8 Results for the United States: Prediction based on the SEIR model considering the variation of the parameter R_0. Figures **a**, **b** and **c** $\gamma = 0.2$ and $R_0 = [2.5, 3.0, 3.9]$, respectively

behavior but also by the containment measures established by the local Governments. For example, if a social-distancing campaign with strict circulation rules is success-fully implemented, this will certainly impact the selection of the R_0 and consequent infection rates in that country.

Since the behavior of the curves tends to be dynamic during a pandemic, a variation of the SEIR model with an intervention factor will be considered in the following section.

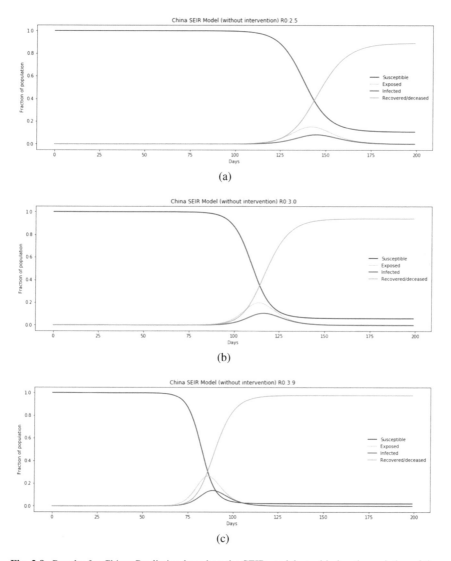

Fig. 2.9 Results for China: Prediction based on the SEIR model considering the variation of the parameter R_0. Figures **a**, **b** and **c** $\gamma = 0.2$ and $R_0 = [2.5, 3.0, 3.9]$, respectively

2.3.3 SEIR Model with Intervention

The application of standard compartmental models is a valid approach for the overall comprehension of an epidemic behavior, even if, depending on the parameters selection, the exponential trend is apparently too high. One limitation of these models is that the parameters are static, which does not reflect any internal or external change

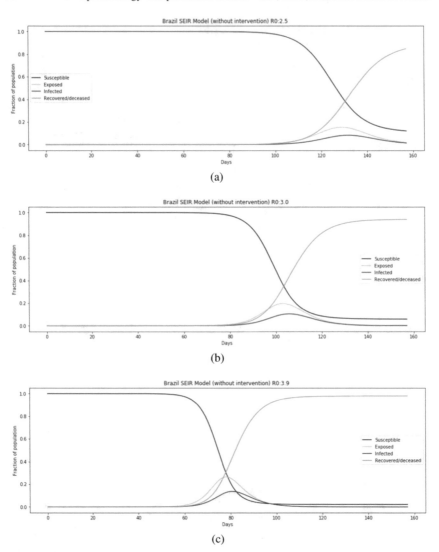

Fig. 2.10 Results for Brazil: Prediction based on the SEIR model considering the variation of the parameter R_0. Figures **a**, **b** and **c** $\gamma = 0.2$ and $R_0 = [2.5, 3.0, 3.9]$, respectively

during the epidemic. This may lead to data misinterpretation, especially considering the SIR model.

In this section, the results of the modified SEIR model with the introduction of an intervention rate L are presented. This modification will change the rate of infection R_0, reflecting possible control measures, such as social isolation. Specifically focusing on the COVID-19 analysis at the present moment, each country under analysis

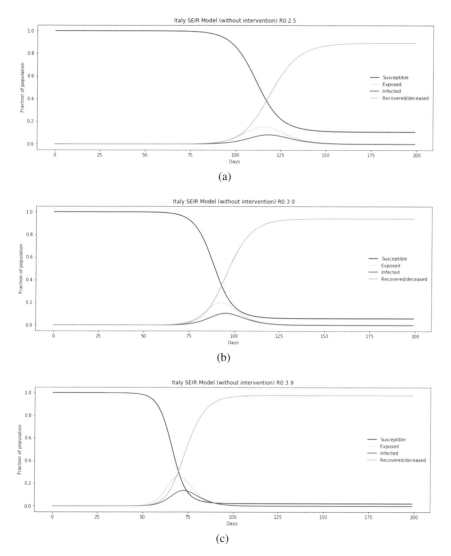

Fig. 2.11 Results for Italy: Prediction based on the SEIR model considering the variation of the parameter R_0. Figures **a**, **b** and **c** $\gamma = 0.2$ and $R_0 = [2.5, 3.0, 3.9]$, respectively

is following a different strategy to contain the virus and the application of any kind of intervention rate should reflect this.

Some graphical interpretation can be performed when analyzing the three plots in parallel to support decision-makers. For example, if one country successfully establishes a social-distancing program, the L parameter shall decrease and this will reduce the peak of the number of infected and exposed curves and will converge

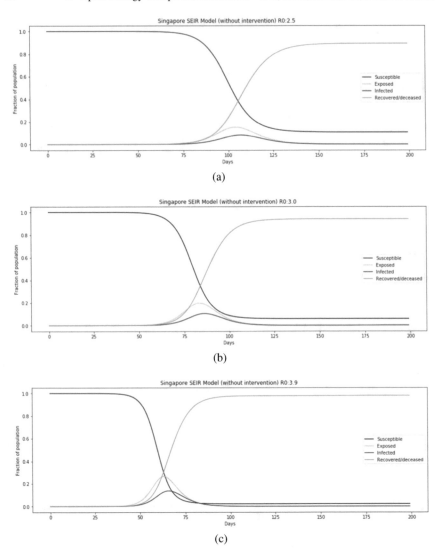

Fig. 2.12 Results for Singapore: Prediction based on the SEIR model considering the variation of the parameter R_0. Figures **a**, **b** and **c** $\gamma = 0.2$ and $R_0 = [2.5, 3.0, 3.9]$, respectively

the susceptible to a stabilization sooner, meaning that the measures didn't allow the virus to spread among the whole population, as can be seen in Fig. 2.18.

In Fig. 2.18a, considering $L = 150$, the peak in the number of infections was located around the center between 75 and 100 days, the number of susceptible decreased to less than 10% of the fraction of population, and the number of recovered increased to more than 90% as well. The implementation of new measures of contention of the virus spread would change the L and lead to a result similar

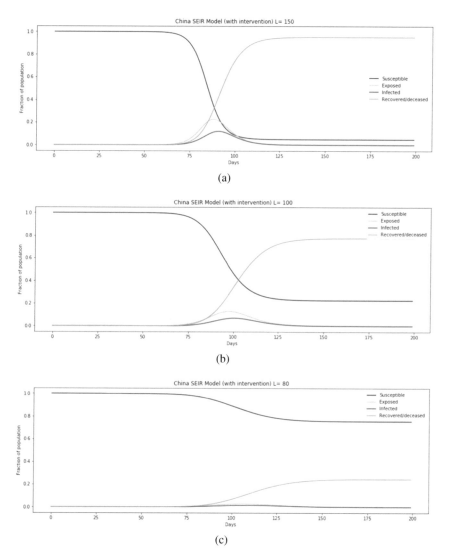

Fig. 2.13 Results for China: Prediction based on the SEIR model with intervention rate L. Figures **a**, **b**, and **c** considering $R_0 = 3.9$ and $L = [150, 100, 80]$, respectively

to Fig. 2.18b, which considers $L = 100$. This change reduced the peak in both the number of infections and exposed subjects, and delayed this peak to almost 100 days, the number of susceptible reduced to become stable around 20% of the fraction of the population and the number of recovered were stable around 80%. Finally, eventually implementing more aggressive contention measures may lead to the analysis presented in Fig. 2.18c, considering $L = 80$. For this prediction, the peak in the number of infections occurs only after 100 days, the number of susceptible subjects was

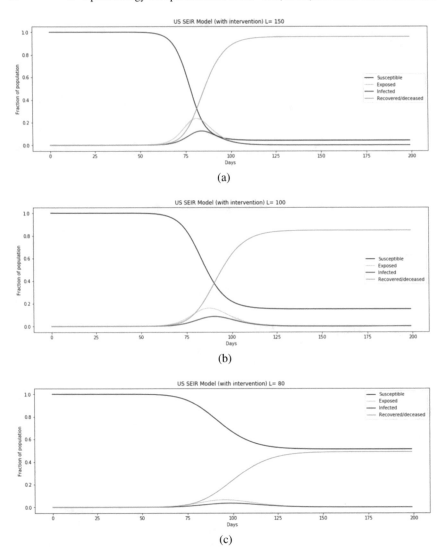

Fig. 2.14 Results for the United States: Prediction based on the SEIR model with intervention rate L. Figures **a**, **b**, and **c** considering $R_0 = 3.9$ and $L = [150, 100, 80]$, respectively

stabilized in around 80% and the number of recovered was kept in only around 20% of the fraction of population.

To comprehend the impact of decision-makers on the epidemic spread, the case of the City of Wuhan, in China, the first place where the COVID-19 epidemic started to take larger proportions is described in detail by Hou et al. [6]. In this paper, the authors explored the effectiveness of the quarantine of the Wuhan city against this epidemic, transmission dynamics of COVID-19, considering an adapted SEIR compartmental

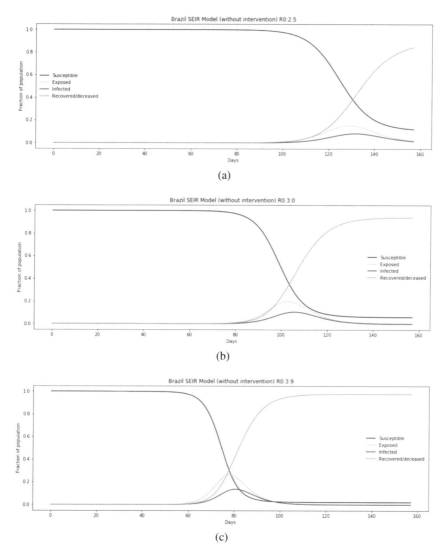

Fig. 2.15 Results for Brazil: Prediction based on the SEIR model with intervention rate L. Figures **a**, **b**, and **c** considering $R_0 = 3.9$ and $L = [150, 100, 80]$, respectively

model based on the epidemiological characteristics of individuals, clinical progression of COVID-19, and quarantine intervention measures of the authority. Some assumptions were considered such as: infected individuals as contagious during the latency period, contact rate of latent individuals were considered in the interval 6–18, and used as a reference for setting the quarantine and isolation interventions. The results presented a very rich analysis with several possibilities and clearly show that, by reducing the contact rate of latent individuals, interventions such as quarantine and

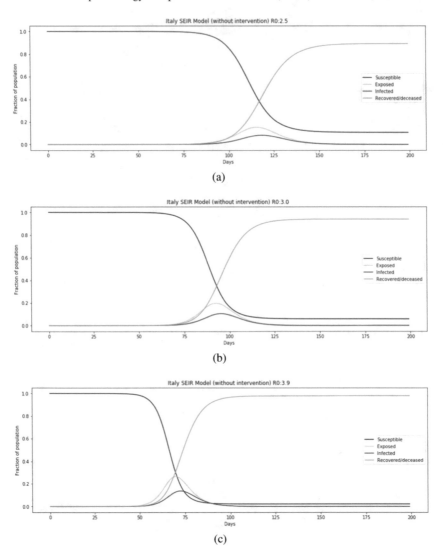

Fig. 2.16 Results for Italy: Prediction based on the SEIR model with intervention rate L. Figures **a**, **b**, and **c** considering $R_0 = 3.9$ and $L = [150, 100, 80]$, respectively

isolation can effectively reduce the potential peak number of COVID-19 infections and delay the time of peak infection.

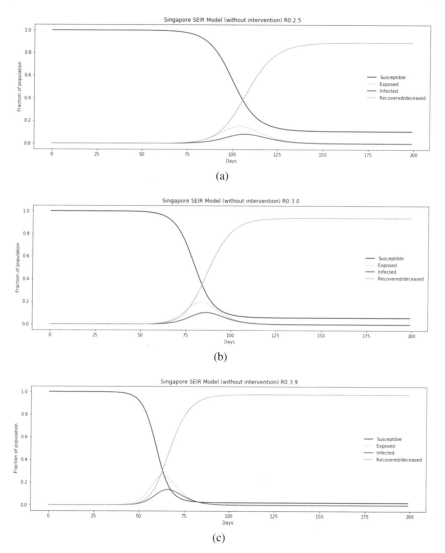

Fig. 2.17 Results for Singapore: Prediction based on the SEIR model with intervention rate L. Figures **a**, **b**, and **c** considering $R_0 = 3.9$ and $L = [150, 100, 80]$, respectively

2.4 Conclusion

The application of compartmental epidemiological models is widely popular during the COVID-19 pandemic but many predictions weren't verified because the modeling could not represent the real models, which were dependent on several external factors and measures of infection contention defined by public health managers. As expected, when using these models, the results presented for both the SIR and SEIR models

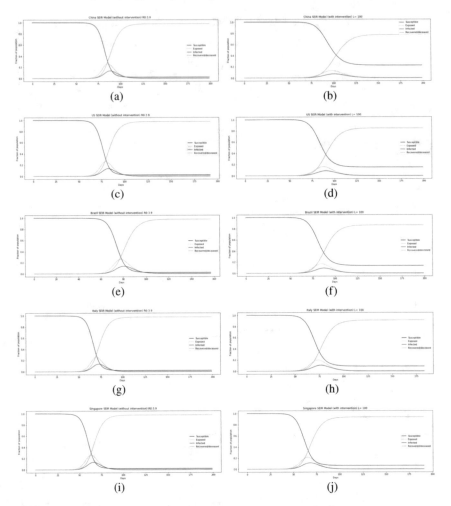

Fig. 2.18 Comparison of SEIR and SEIR with intervention models for a similar $R_0 = 3.9$, considering the intervention rate $L = 150$. Results for China **a** and **b**; United States **c** and **d**; Brazil (**e**) and (**f**); Italy (**g**) and (**h**) and Singapore (**i**) and (**j**)

were strongly dependent on the parameters selection. Each parameter is responsible for the rate of transitions between one compartment and the next one.

In the case of SIR models, for example, at the beginning of the pandemic in some countries, several publications and articles were considering the model as ground truth and assuring that the pandemic would peak and reduce to near zero exponentially in just a few days and this was not confirmed in real cases.

One important result could be achieved when considering the SEIR with Intervention model, since it reflects in terms of infection rates the containment measures adopted by the governments.

Compartmental models are valid approaches for comprehending and analyzing epidemiological data, especially if the model is adjusted to consider specific aspects of the epidemic under analysis, as in the case of the COVID-19 pandemic.

The following chapter will present the technical aspects and the results of COVID-19 pandemic forecasting based on the linear predictor called Autoregressive Integrated Moving Average (ARIMA).

References

1. N. Bacaër, N. Bacaër, Daniel Bernoulli, d'Alembert and the inoculation of smallpox (1760), in *A Short History of Mathematical Population Dynamics* (Springer, London, 2011), pp. 21–30. https://doi.org/10.1007/978-0-85729-115-8_4. https://link.springer.com/chapter/10.1007/978-0-85729-115-8

2. K.L. Cooke, P. Van Den Driessche, Analysis of an SEIRS epidemic model with two delays. J. Math. Biol. **35**(2), 240-260 (1996). issn: 14321416. https://doi.org/10.1007/s002850050051. https://link.springer.com/article/10.1007/s002850050051

3. G. Ellison, Implications of heterogeneous SIR models for analyses of COVID-19. Tech. rep. (National Bureau of Economic Research, Cambridge, MA, 2020). https://doi.org/10.3386/w27373. http://www.nber.org/papers/w27373.pdf

4. M. Gatto et al., Spread and dynamics of the COVID-19 epidemic in Italy: Effects of emergency containment measures. Proc. Natl. Acad. Sci. U. S. A. **117**(19), 10484–10491 (2020). issn: 10916490. https://doi.org/10.1073/pnas.2004978117

5. G. Giordano et al., Modelling the COVID-19 epidemic and implementation of populationwide interventions in Italy. Nat. Med. **26**(6), 855–860 (2020). issn: 1546170X. https://doi.org/10.1038/s41591-020-0883-7

6. C. Hou et al., The effectiveness of quarantine of Wuhan city against the Corona Virus Disease 2019 (COVID-19): a well-mixed SEIR model analysis. J. Med. Virol. **92**(7), 841–848 (2020). issn: 0146-6615. https://doi.org/10.1002/jmv.25827. https://onlinelibrary.wiley.com/doi/abs/10.1002/jmv.25827

7. W.O.K. McKendrick, A.G. McKendrick, A contribution to the mathematical theory of epidemics. Proc. R. Soc. Lond. Ser. A, Contain. Pap. Math. Phys. Character **115**(772), 700–721 (1927). issn: 0950-1207. https://doi.org/10.1098/rspa.1927.0118

8. T. Rhodes, K. Lancaster, M. Rosengarten, A model society: maths, models and expertise in viral outbreaks (2020). https://doi.org/10.1080/09581596.2020.1748310

9. P. Samui, J. Mondal, S. Khajanchi, A mathematical model for COVID-19 transmission dynamics with a case study of India. Chaos Solitons Fractals, p. 110173 (2020). issn: 09600779. https://doi.org/10.1016/j.chaos.2020.110173. https://linkinghub.elsevier.com/retrieve/pii/S0960077920305695

10. J.T. Wu, K. Leung, G.M. Leung, Nowcasting and forecasting the potential domestic and international spread of the 2019-nCoV outbreak originating in Wuhan, China: a modelling study. Lancet **395**(10225), 689–697 (2020). issn: 1474547X. https://doi.org/10.1016/S0140-6736(20)30260-9

11. X. Zhou et al., Forecasting the worldwide spread of COVID-19 based on logistic model and SEIR model. medRxiv, p. 2020.03.26.20044289 (2020). https://doi.org/10.1101/2020.03.26.20044289. http://medrxiv.org/lookup/doi/10.1101/2020.03.26.20044289

Chapter 3
Forecasting COVID-19 Time Series Based on an Autoregressive Model

3.1 Introduction

A time series is expressed as a set of data points arranged in time and its analysis intends to reveal reliable and meaningful statistics that can be used to interpret information and possibly forecast future values. ARIMA, Autoregressive Integrated Moving Average, was introduced by Box and Jenkins in the 1970s and presented in [2]. This model takes into consideration changing disturbances in time and tendencies [3].

The approach assumes characteristics of stationarity and seasonality for the considered time series, and specified by autoregressive tendency differences, and moving average operators. These parameters are used to define the order of the model, and this is normally defined by grid search or based on the analysis of autocorrelation (ACF) and partial autocorrelation (PACF) functions [7].

Several studies already considered the application of the ARIMA model to predict COVID-19. These studies indicate that ARIMA models are acceptable for predicting COVID-19's incidence later on. The investigation results can shed light on understanding this outbreak's trends and provide an idea of this epidemiological stage of those regions.

Benvenuto et al. [1] used ARIMA to determine the incidence of COVID-19. The study selects $ARIMA(1, 0, 4)$ as the best fit model. This work performed a statistical investigation on the prevalence and incidence datasets. The analysis of ACF and PACF correlogram showed both the incidence and prevalence of COVID-19 are not affected by the seasonality.

Ceylan [3] estimates the prevalence of COVID-19 in Italy, Spain, and France. The data analyzed in this study corresponds to the period from February to April 2020. Several ARIMA models were formulated with different ARIMA parameters. After comparison, $ARIMA(0, 2, 1)$, $ARIMA(1, 2, 0)$, and $ARIMA(0, 2, 1)$ models with the lowest error values (4.7520, 5.8486, and 5.6335) were selected as the best models

© The Author(s), under exclusive license to Springer Nature Switzerland AG 2021
J. A. L. Marques et al., *Predictive Models for Decision Support in the COVID-19 Crisis*, SpringerBriefs in Applied Sciences and Technology, https://doi.org/10.1007/978-3-030-61913-8_3

for Italy, Spain, and France, respectively, showing that different models shall be designed for each country under analysis.

The main objective of this chapter is to evaluate the results of the application of the ARIMA regression model on the prediction of epidemiological time series during the COVID-19 pandemic, based on the evaluation of the R^2 Score and the error based measures MSE and MAE.

3.2 Materials and Methods

As stated in its acronym, the ARIMA model integrates an Autoregressive (AR) and Moving Average (MA) models and some steps shall be performed before defining the best fit for the model. First of all, time-series stationarity and seasonality need to be verified. There are significant drawbacks to this kind of assumption when considering real-life time series, such as the epidemiological curves for the COVID-19 pandemic. Since the number of samples is still limited and new data are being continuously generated.

Nevertheless, these requirements can be approximately assumed applying numerical techniques such as the Augmented Dickey–Fuller (ADF) unit-root evaluation, which helps to estimate whether the time series is stationary. Log transformation and differences are the preferred approaches to stabilize the time series. Finally, seasonal and nonseasonal differences can be used to stabilize trends and periodicity [1].

After verifying the time series requirements, Autocorrelation Function (ACF) graph and Partial Autocorrelation Correlogram (PACF) are reliable approaches to estimate the parameters of the ARIMA model. The ARIMA model parameters are described as follows:

- p, the number of lag observations included in the design also called the lag order;
- d, the number of times that the raw observations are differenced;
- q, the moving average window size.

As previously discussed, the ARIMA modeling procedure consists of four iterative steps: assessment of the design, estimation of parameters, diagnostic checking, and forecast. The first thing of the ARIMA model is to control if the time series is seasonal and stationary. When its statistical properties such as mean, variance, autocorrelation are continuous in time, a time series can be considered as stationary. The ACF graph determines whether preceding values in the series are linked to the following values, while the PACF graph finds out the amount of correlation between a factor, and a consequence of this stated factor that's not explained by correlation at all non-order lags [3].

3.2.1 Proposed Methodology for the Data Analysis

A general description of the considered methodology for this chapter, Chaps. 4 and 5 of the book is presented in Chap. 1, Sect. 1.3. In that section are presented the source of COVID-19 related data; the justification of considering five countries on the analysis; the data handling approach (k-fold cross-validation strategy) and, finally, the evaluation criteria, considering R2 Score, Mean Square Error (MSE), and Mean Absolute Error (MAE). Essential references for each topic are also presented there.

The development environment was structured as a Python notebook [6], hosted by Kaggle [4], including data analysis and manipulation packages, such as NumPy and Pandas [5] and Statsmodel [8].

The approximation technique considered for the definition of the three parameters (p, q, and d) of the ARIMA model was the grid search (Grid Search CV), which tests a wide set of parameter tuning executions and selects the best fit. The results were organized in the following sequence of countries: China, USA, Brazil, Italy, and Singapore.

3.3 Results and Discussion

In this chapter, the results are presented according to the following steps. For each country, a figure with five time series selected during the cross-validation phase is presented with their corresponding prediction. The three evaluation criteria, R^2, MSE, and MAE, are discussed and compared.

3.3.1 ARIMA Predictions for China

According to the method described in Sect. 3.2, the best time-series prediction performance for China was achieved by the $ARIMA(p = 1, q = 2, d = 0)$ model.

In Fig. 3.1, are presented 5 time series of active cases in China, considered as testing samples and the corresponding prediction for each of them.

The model obtained an average R^2 score metric of 0.840787 and an average MAE of 57.21552. For the MSE, since it is a quadratic measure, the average doesn't make really sense to consider, because the penalties for larger individual errors are emphasized. Since the R^2 Score is a percentage measure, as previously explained in Chap. 1, when analyzing each prediction it is possible to see that the best results, closer to 1.0 (100%) are found for the plots where the time-series behavior is smooth and with no significant changes. The best R^2 Score performance is 99.7228%, achieved for Sample 1, presented in Fig. 3.1a, while the worse result is 27.8608% for Sample 4, which is a time series with several short term variations.

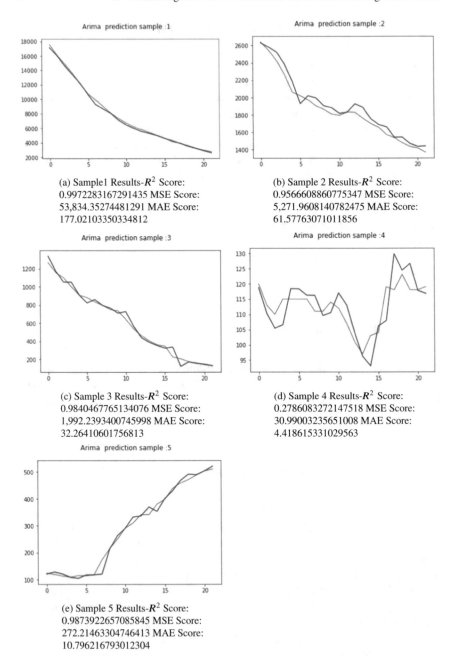

(a) Sample1 Results-R^2 Score: 0.9972283167291435 MSE Score: 53,834.35274481291 MAE Score: 177.02103350334812

(b) Sample 2 Results-R^2 Score: 0.9566608860775347 MSE Score: 5,271.9608140782475 MAE Score: 61.57763071011856

(c) Sample 3 Results-R^2 Score: 0.9840467765134076 MSE Score: 1,992.2393400745998 MAE Score: 32.26410601756813

(d) Sample 4 Results-R^2 Score: 0.2786083272147518 MSE Score: 30.99003235651008 MAE Score: 4.418615331029563

(e) Sample 5 Results-R^2 Score: 0.9873922657085845 MSE Score: 272.21463304746413 MAE Score: 10.796216793012304

Fig. 3.1 Prediction Results of the ARIMA($p = 1$, $q = 2$, $d = 0$) model for 5 samples of COVID-19 Active Cases time series from China. The blue lines are the original data, while the red lines are the prediction time series

According to the results, the lowest MAE Score was 4.4186 for Sample 4, while the highest MAE Score was 177.0210 for Sample 1. The same for MSE, Sample 4 presented the lowest quadratic error, 30.9900, while Sample 1 MSE was 53, 834.3527.

Analyzing the performance of the distance-based error metrics, MSE and MAE Scores, it is important to consider that their results are dependent on the sample values, since the distance is calculated based on the subtraction of the average. That's the reason why Sample 4, which achieved the worse performance for R^2 Score, would be considered the best result, since it presents the lowest values for SME and SAE.

The reader must carefully notice that the sample values are oscillating from 95 to 130, while on Sample 3, the range is between 200 and 1,200 and on Sample 1, from 2,000 to 18,000. This leads us to be careful about the analysis of the SME and SAE to compare the results between samples of different ranges. The variance is an auxiliary statistical metric that can be used to support the analysis. In this case, we consider the average of the MAE Score as an acceptable Score, since the accumulated error is a more suitable metric than the quadratic error for the analysis.

3.3.2 ARIMA Predictions for the United States of America

A brief introduction about the pandemic situation in the United States of America and the epidemiological curve of the number of confirmed new cases per day was presented in Chap. 1. These aspects will reflect on the prediction and are important to be understood.

According to the method described in Sect. 3.2, the best time-series prediction performance for the USA was achieved by the $ARIMA(p = 2, q = 1, d = 0)$ model.

In Fig. 3.2, are presented 5 time series of active cases in the USA, considered as testing samples and the corresponding prediction for each of them.

The model obtained an average R^2 score metric of 0.99 and an average accumulated error MAE of 8, 261. For the MSE, again since it is a quadratic measure, the average doesn't make really sense to consider, because the penalties for larger individual errors are emphasized.

Since the R^2 Score is a percentage measure, as previously explained in Chap. 1, when analyzing each prediction it is possible to see that the best results, closer to 1.0 (100%) are found for the plots where the time-series behavior is smooth and with no significant changes. The best R^2 Score performance is 99.9152%, achieved for Sample 2, presented in Fig. 3.2b, while the worse result is 94.6761% for Sample 4, which is a time series with an increasing trend with two intermediary oscillations.

According to the results, the lowest MAE Score was 1, 740.8037 for Sample 1, while the highest MAE Score was 8, 302.5034 for Sample 4. The same for MSE, Sample 1 presented the lowest quadratic error, while Sample 4 presented the highest MSE.

Analyzing the performance of the distance-based error metrics, MSE and MAE Scores, it is important to consider that their results are dependent on the sample values, since the distance is calculated based on the subtraction of the average. For

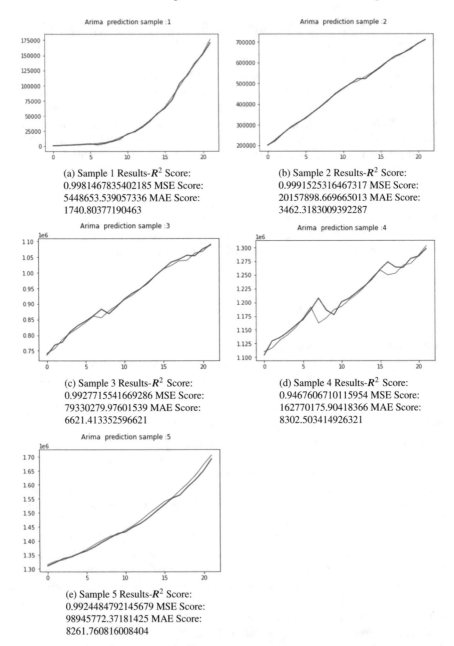

(a) Sample 1 Results-R^2 Score: 0.9981467835402185 MSE Score: 5448653.539057336 MAE Score: 1740.80377190463

(b) Sample 2 Results-R^2 Score: 0.9991525316467317 MSE Score: 20157898.669665013 MAE Score: 3462.3183009392287

(c) Sample 3 Results-R^2 Score: 0.9927715541669286 MSE Score: 79330279.97601539 MAE Score: 6621.413352596621

(d) Sample 4 Results-R^2 Score: 0.9467606710115954 MSE Score: 162770175.90418366 MAE Score: 8302.503414926321

(e) Sample 5 Results-R^2 Score: 0.9924484792145679 MSE Score: 98945772.37181425 MAE Score: 8261.760816008404

Fig. 3.2   Prediction Results of the ARIMA($p = 2, q = 1, d = 0$) model for 5 samples of COVID-19 Active Cases time series from the United States. The blue lines are the original data, while the red lines are the prediction time series

the present analysis, the US data series presented usually a large sample range, denoting the large increase in the number of active cases in that country. This reflects directly on the absolute values of SME and SAE. The reader must carefully notice that the sample values are oscillating from close to 0 to 175,000 on Sample 1 and from 1.30×10^6 to 1.70×10^6 on Sample 5. This leads us to be careful about the analysis of the SME and SAE to compare the results between samples of different ranges. The variance is an auxiliary statistical metric that can be used to support the analysis. In this case, we consider the average of the MAE Score as an acceptable Score, since the accumulated error is a more suitable metric than the quadratic error for the analysis.

3.3.3 ARIMA Predictions for Brazil

A brief introduction about the pandemic situation in Brazil and the epidemiological curve of the number of confirmed new cases per day was presented in Chap. 1. These aspects will reflect on the prediction and are important to be understood.

According to the method described in Sect. 3.2, the best time-series prediction performance for Brazil was achieved by the $ARIMA(p = 0, q = 1, d = 1)$ model. In Fig. 3.3, are presented 5 time series of active cases in Brazil, considered as testing samples and the corresponding prediction for each of them.

Again, analyzing the average performance for this country, the model obtained an average R^2 score metric of 0.88 and an average accumulated error MAE of 3, 824. The same comments presented previously for the MSE, again since it is a quadratic measure, the average doesn't make really sense to consider, because the penalties for larger individual errors are emphasized. For Brazil, the best R^2 Score performance is 99.4958%, achieved for Sample 4, presented in Fig. 3.3d, which is a sample of a quasilinear behavior. Since the R^2 Score is a percentage measure, as previously explained in Chap. 1, when analyzing each prediction it is possible to see that the best results, closer to 1.0 (100%), are found for the plots where the time-series behavior is smooth and with no significant changes. The worse result is 57.2268% for Sample 2, which is a time series with one significant oscillation.

According to the results, the lowest MAE Score was 156.4406 for Sample 1, while the highest MAE Score was 10, 804.0219 for Sample 5. The same for MSE, Sample 1 presented the lowest quadratic error, while Sample 5 presented the highest MSE. A visual inspection of Sample 5 may indicate that the error was significantly increased because of the oscillation in the first group of samples.

The same considerations about the dependency of the sample range for the analysis of MSE and MAE Scores apply here. Please check the comments from the previous subsections.

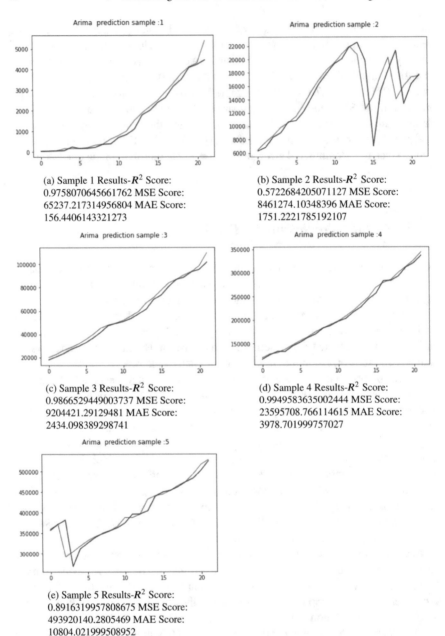

(a) Sample 1 Results-R^2 Score: 0.9758070645661762 MSE Score: 65237.217314956804 MAE Score: 156.4406143321273

(b) Sample 2 Results-R^2 Score: 0.5722684205071127 MSE Score: 8461274.10348396 MAE Score: 1751.2221785192107

(c) Sample 3 Results-R^2 Score: 0.9866529449003737 MSE Score: 9204421.29129481 MAE Score: 2434.098389298741

(d) Sample 4 Results-R^2 Score: 0.9949583635002444 MSE Score: 23595708.766114615 MAE Score: 3978.701999757027

(e) Sample 5 Results-R^2 Score: 0.8916319957808675 MSE Score: 493920140.2805469 MAE Score: 10804.021999508952

Fig. 3.3 Prediction Results of the ARIMA($p = 0, q = 1, d = 1$) model for 5 samples of COVID-19 Active Cases time series from Brazil. The blue lines are the original data, while the red lines are the prediction time series

3.3.4 ARIMA Predictions for Italy

A brief introduction about the pandemic situation in Italy and the epidemiological curve of the number of confirmed new cases per day was presented in Chap. 5. These aspects will reflect on the prediction and are important to be understood.

According to the method described in Sect. 3.2, the best time-series prediction performance for Italy was achieved by the $ARIMA(p = 2, q = 1, d = 0)$ model. In Fig. 3.4, are presented 5 time series of active cases in Italy, considered as testing samples and the corresponding prediction for each of them.

Analyzing the average performance of the model for this country, it was obtained an average R^2 score metric of 0.98 and an average accumulated error MAE of 724. The same comments presented previously about the MSE apply here. Since it is a quadratic measure, the average doesn't make really sense to consider, because the penalties for larger individual errors are emphasized.

For Italy, the best R^2 Score performance is 99.8301%, achieved for Sample 2, presented in Fig. 3.4b, which is a sample of smooth increasing behavior. Since the R^2 Score is a percentage measure, as previously explained in Chap. 1, when analyzing each prediction it is possible to see that the best results, closer to 1.0 (100%), are found for the plots where the time series behavior is smooth and with no significant changes. The worse result is 97.5807% for Sample 1, which is also a very positive result.

According to the results, the lowest MAE Score was 151.2406 for Sample 1, while the highest MAE Score was 740.8156 for Sample 2. The same for MSE, Sample 1 presented the lowest quadratic error, while Sample 2 presented the highest MSE. It is important to notice that all the Scores presented good results for the 5 time series. This is probably because the epidemiological curve in Italy presented an expected behavior, with exponential growth, well-delimited peak, and consequent reduction after the contention measures.

The same considerations about the dependency of the sample range for the analysis of MSE and MAE Scores apply here. Please check the comments from the first subsection.

3.3.5 ARIMA Predictions for Singapore

A brief introduction about the pandemic situation in Singapore and the epidemiological curve of the number of confirmed new cases per day was presented in Chap. 1. These aspects will reflect on the prediction and are important to be understood.

According to the method described in Sect. 3.2, the best time-series prediction performance for Singapore was achieved by the $ARIMA(p = 2, q = 2, d = 0)$ model. In Fig. 3.5, are presented 5 time series of active cases in Singapore, considered as testing samples and the corresponding prediction for each of them.

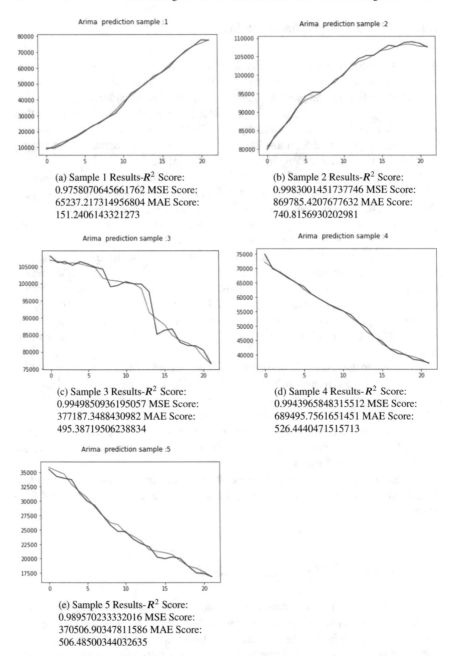

(a) Sample 1 Results-R^2 Score: 0.9758070645661762 MSE Score: 65237.217314956804 MAE Score: 151.2406143321273

(b) Sample 2 Results-R^2 Score: 0.9983001451737746 MSE Score: 869785.4207677632 MAE Score: 740.8156930202981

(c) Sample 3 Results-R^2 Score: 0.9949850936195057 MSE Score: 377187.3488430982 MAE Score: 495.38719506238834

(d) Sample 4 Results-R^2 Score: 0.9943965848315512 MSE Score: 689495.7561651451 MAE Score: 526.4440471515713

(e) Sample 5 Results-R^2 Score: 0.989570233332016 MSE Score: 370506.90347811586 MAE Score: 506.48500344032635

Fig. 3.4 Prediction Results of the ARIMA($p = 2, q = 1, d = 0$) model for 5 samples of COVID-19 Active Cases time series from Italy. The blue lines are the original data, while the red lines are the prediction time series

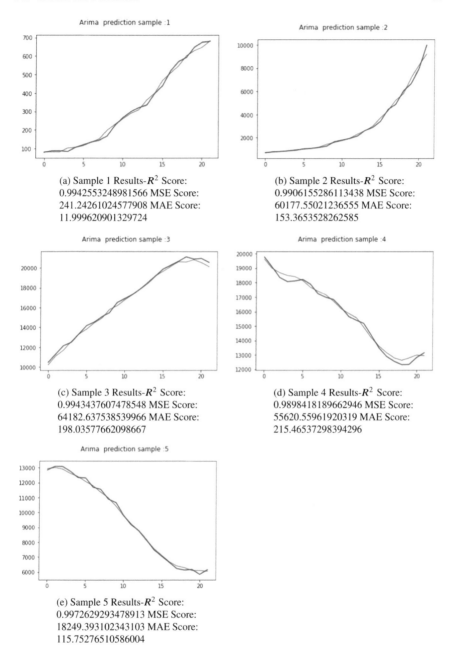

(a) Sample 1 Results-R^2 Score: 0.9942553248981566 MSE Score: 241.24261024577908 MAE Score: 11.999620901329724

(b) Sample 2 Results-R^2 Score: 0.9906155286113438 MSE Score: 60177.55021236555 MAE Score: 153.3653528262585

(c) Sample 3 Results-R^2 Score: 0.9943437607478548 MSE Score: 64182.637538539966 MAE Score: 198.03577662098667

(d) Sample 4 Results-R^2 Score: 0.9898418189662946 MSE Score: 55620.55961920319 MAE Score: 215.46537298394296

(e) Sample 5 Results-R^2 Score: 0.9972629293478913 MSE Score: 18249.393102343103 MAE Score: 115.75276510586004

Fig. 3.5 Prediction Results of the ARIMA(p = 2, q = 2, d = 0) model for 5 samples of COVID-19 Active Cases time series from Singapore. The blue lines are the original data, while the red lines are the prediction time series

Analyzing the average performance of the model for this country, it was obtained an average R^2 score metric of 0.99 and an average accumulated error MAE of 138. The same comments presented previously about the MSE apply here. Since it is a quadratic measure, the average doesn't make really sense to consider, because the penalties for larger individual errors are emphasized.

The best R^2 Score performance for Singapore was 99.7262%, achieved for Sample 5, presented in Fig. 3.5e, which is a sample of smooth decreasing behavior. Since the R^2 Score is a percentage measure, as previously explained in Chap. 1, when analyzing each prediction it is possible to see that the best results, closer to 1.0 (100%), are found for the plots where the time series behavior is smooth and with no significant changes. The worse result is 98.9841% for Sample 4, which is also a very positive result.

According to the results, the lowest MAE Score was 11.9996 for Sample 1, while the highest MAE Score was 215.4653 for Sample 4. The same for MSE, Sample 1 presented the lowest quadratic error, while Sample 4 presented the highest MSE. It is important to notice that all the Scores presented good results for the 5 time series. This is probably because the absolute number of cases in Singapore never increased significantly and the epidemiological curve in Singapore presented an expected behavior, with exponential growth, well-delimited peak, and consequent reduction after the contention measures, despite the fact that a few oscillations of low amplitude are still present.

The same considerations about the dependency of the sample range for the analysis of MSE and MAE Scores apply here. Please check the comments from the first subsection.

3.3.6 Model Performance Comparison Between Countries

The last analysis that may be considered is the comparison of the best ARIMA models between the five different countries considered in this work. The results, including the best ARIMA model and its parameters, are presented in Table 3.1.

The results of R^2 Score are presented in Table 3.2. It is important to notice that, for this score, the highest the results, the better, since this is a ratio, i.e., a measure based on a percentage. On the other hand, the best and worst results of the MAE Score are in Table 3.3. This metric is a measure of a sum of differences for each sample, so the lowest the result, the better, with a lower accumulated error.

3.4   Conclusion

According to the presented discussion, autoregressive models integrated with moving average can be considered as a prediction model for the COVID-19 epidemiological data, with certain limitations being considered. Results of 5 different countries were

Table 3.1 Compilation of the best parameters for the ARIMA autoregressive model with the average of R^2 Score and MAE Score for the 5 time series used for validation of each country

Country	Best model	Average R^2 score	Average MAE score
China	ARIMA(1, 2, 0)	0.84	57.21
US	ARIMA(2, 1, 0)	0.99	8,26.14
Brazil	ARIMA(0, 1, 1)	0.88	3,824.02
Italy	ARIMA(2, 1, 0)	0.98	724.83
Singapore	ARIMA(2, 2, 0)	0.99	138.62

Table 3.2 Highest (best) and Lowest (worst) results of the R^2 Score for the ARIMA auto regressive model considering the 5 time series used for validation of each country

Country	Highest R^2 score	Lowest R^2 score
China	0.9972	0.2786
US	0.9991	0.9467
Brazil	0.9949	0.5722
Italy	0.9983	0.9758
Singapore	0.9972	0.9898

Table 3.3 Lowest (best) and Highest (worst) results of the MAE Score for the ARIMA auto regressive model considering the 5 time series used for validation of each country

Country	Lowest MAE	Highest MAE
China	4.41	177.02
US	1,740.80	8,302.50
Brazil	156.44	10,804.02
Italy	151.24	740,81
Singapore	11.99	215.46

presented with R^2 Score ranging from 0.84 when predicting the number of infected from China to 0.99 for the US and Singapore. On the other hand, the accumulated error (MAE) from China was the lowest, 57.21, while for the US it was the highest, 8, 261.

Higher R^2 Scores were obtained when the sample time series was smoothly increasing or decreasing. The error metrics were higher when the prediction was performed for oscillating data series. This may indicate that the use of ARIMA models may be suitable as a prediction tool for the COVID-19 when the country is not facing severe oscillations in the number of infections. For time series with constant changes in the epidemiological curves, the errors tend to significantly grow.

It is important to notice that the ARIMA model assumes requirements of stationarity and seasonality that may be determined approximately by numerical calculations but are often considered weak assumptions. Another key aspect is related to the definition of the model parameters. A detailed analysis of the ACF or PACF behaviors should be carefully conducted to characterize the time series under analysis.

Linear autoregressive analysis is considered in every prediction application as a first step tool not only because of the simplicity for the implementation, but also because the results are usually satisfactory for a large number of applications. Other techniques based on Kalman Quadratic Filter and AI LSTM neural networks are presented in the following chapters and a broader view about prediction techniques will be possible for the reader.

The following chapter will focus on a new strategy of prediction, based on a quadratic Kalman filter predictor.

References

1. D. Benvenuto et al., Application of the ARIMA model on the COVID-2019 epidemic dataset. Data in Brief **29**, 105340 (2020), issn: 23523409. https://doi.org/10.1016/j.dib.2020.105340.
2. G. Box et al., *Time Series Analysis: Forecasting and Control*, 5th edn (Wiley, Hoboken, 2006). https://www.wiley.com/en-us/Time+Series+Analysis7B5C7D3A+Forecasting+and+Control7 B5C7D2C+5th+Edition-p-9781118675021 (visited on 08/04/2020)
3. Z. Ceylan, Estimation of COVID-19 prevalence in Italy, Spain, and France. Sci. Total Environ. **729**, 138817 (2020). issn: 18791026. https://doi.org/10.1016/j.scitotenv.2020.138817.
4. Kaggle, Kaggle: your home for data science (2020). https://www.kaggle.com/ (visited on 08/04/2020)
5. Pandas, Pandas - python data analysis library (2020). https://pandas.pydata.org/ (visited on 08/04/2020)
6. Python, Welcome to python.org (2020). https://www.python.org/ (visited on 08/04/2020)
7. M.H.D.M. Ribeiro et al., Short-termforecasting COVID-19 cumulative confirmed cases: perspectives for Brazil. Chaos, Solitons Fractals **135** (2020), issn: 09600779. https://doi.org/10. 1016/j.chaos.2020.109853
8. Statsmodel, Statsmodels - python package (2020). https://www.statsmodels.org/stable/index. html (visited on 08/04/2020)

Chapter 4
Nonlinear Prediction for the COVID-19 Data Based on Quadratic Kalman Filtering

4.1 Introduction

The Kalman Filter is a state-space model that is used in several applications as a predictor. The filter algorithm requires low computational power and provides estimates of some unknown variables given the measurements observed over time. The mathematical concepts are not so simple to understand if the reader has no familiarity with estimation theory, for example.

In practice, the method considers a set of measures observed over an interval, including noise, and estimates new samples, according to the considered time series or variable. The first concept is to understand that it considers a joint probability distribution across the variables for each time frame. To simplify, the Kalman Filter (KF) is an optimization estimator which suggests parameters of interest from previous observations.

The KF aims to find the "most reliable estimate" from noisy input. The filter presents a recursive resolution to the linear optimal filtering problem to stationary as well as nonstationary situations, and treats the new measures as they appear. Only the previous estimate is used for calculation, which reduces the need for saving the whole data from previous iterations [2]. These techniques have found application in various disciplines and, across the past two decades, have been used to contagious infection epidemiology [10]. The results of the application of the Quadratic Kalman Filter (QKF) are presented in this chapter.

4.2 Materials and Methods

An introductory background about the necessary mathematical tools and techniques to understand this prediction approach is presented in the following subsections, from the definition of state-space model to the

4.2.1 Kalman Filter

A more detailed description of the Kalman Filter for prediction is necessary to effectively evaluate its application. First of all, the comprehension of what is a state-space model will be briefly explained for the sake of better comprehension from nonmathematicians.

A State-Space Model (SSM), also known in the technical literature as Hidden Markov Model (HMM), can be defined as a class of probabilistic models that describes the probabilistic dependence between a latent state variable and an observed measurement (Koller and Friedman 2009). The term "state space" originated in the area of control engineering (Kalman 1960) and its most popular application is the Kalman Filter, from the linear approach to other large number of variations, such as quadratic KF. The goal is to analyze dynamical systems, which can be deterministic or stochastic, measured as continuous or discrete through a stochastic process. There is a vast number of applications from computer science to biomedical engineering and epidemiology.

The Kalman filter gives a linear minimum error variance estimate of the state characterized by a state-space model. The filter has the support of leading with noise in the couple, model, and the data. The main goal of the KF is to diminish the mean squared error within the real and measured data. Consequently, it gives the accurate as a possible measure of the data in the mean squared error function. Thought from this fact, it should be plausible to determine that the KF has much in common with the chi-square. The chi-square merit function is typically applied as a model to fit a collection of model variables to a method named least-squares fitting. The KF is usually named as recursive least squares (RLS) [1].

The filter dynamics rises from the constant periods of forecasting and filtering. The change aspects of these periods are determined and translated in Gaussian probability density functions. Following the changes in the system constraints, the Kalman filter dynamics converge to a steady-state filter, and the steady-state gain is inferred. The learning method connected with the filter, which describes the new data conveyed to the state measure by the latter system measure, is presented.

4.2.2 State-Space Derivation

The differential equations of the KF can be incorporated into a state-space component. Let $Y_t, Y_{t-1}, \ldots, Y_1$ denoted the observed values of a feature in time t, t-1, ..., 1. We assume that Y, depends on an unobservable quantity θ, known as system state variables. The goal of Kalman Filter is make inferences of θ. The relation between Y_t and θ is given by a equation [1, 7]:

$$Y_t = F_t \theta_t + v_t \qquad (4.1)$$

where F_t is a known quantity. F_t is the noiseless connection between the t state vector and the measurement vector, and is assumed stationary over time. The observation error v_t is the associated with measurement error [4, 7, 9]. The main difference between KF and conventional linear models is that KF regression coefficients are not constant ant change over time as the system equation:

$$\theta_t = G_t \theta_{t-1} + w_t \tag{4.2}$$

where θ is the state vector at time t; G_t is the state transition matrix of the progress from the position at t-1 to the state at t, and is presumed stationary over time; w_t is the associated white noise with recognize covariance; v_t and the system equation error wt are presumed to be mutually independent random variables, spectrally white, and with normal probability distributions. w_t and v_t are sequences of white, Gaussian noise with zero mean:

$$E[w_t] = E[v_t] = 0, \tag{4.3}$$

The Kalman filter is the filter that gets the least mean square state error estimation. When Y_0 is a Gaussian vector, the state and perceptions noises w_t and v_t are white and Gaussian, and the state and observation dynamics are linear. For the minimization of the MSE to support the optimal filter, it must be plausible to evaluate model errors using Gaussian distributions. The covariances of the noise models are considered stationary in period and are given by

$$Q = E[w_t w_t^T] \tag{4.4}$$

$$R = E[v_t v_t^T] \tag{4.5}$$

The mean squared error is given by

$$P_k = E[e_t e_t^T] = E[(Y_t - \hat{Y}_t)(Y_t - \hat{Y}_t)^T] \tag{4.6}$$

where P is the error covariance matrix at time t. Considering the previous estimation of $\hat{Y}$ is named $\hat{Y}'$, and was obtained by observation of the system. It is welcome to estimate using a write an update equation, mixing the old estimation with new measurement data.

4.2.3 Quadratic Kalman Filter

The original application of the KF is in the form of a linear filtering method but several variations are presented in the literature. Assuming that the COVID-19 epidemiological time series have the presence of nonlinear components, this chapter uses the Quadratic Kalman Filter (QKF), where the transition equations are still linear but the measurement equations are quadratic.

4.2.4   Proposed Methodology for the Data Analysis

A general description of the considered methodology for Chaps. 3–5 of the book is presented in Chap. 1, Sect. 1.3. In that section are presented the source of COVID-19 related data; the justification of considering five countries on the analysis; the data handling approach (k-fold cross-validation strategy) and, finally, the evaluation criteria, considering R2 Score, Mean Square Error (MSE), and Mean Absolute Error (MAE). Essential references for each topic are also presented there.

The development environment was structured as a Python notebook [6], hosted by Kaggle [3], including data analysis and manipulation packages, such as NumPy and Pandas [5] and Statsmodel [8].

The results are presented in the next section, organized in the following sequence of countries: China, USA, Brazil, Italy, and Singapore.

4.3   Results and Discussion

In this chapter, the results are presented according to the following steps. For each country, a figure with 5 time series selected during the cross-validation phase is presented with their corresponding prediction. The three evaluation criteria, R^2, MSE, and MAE, are discussed and compared.

4.3.1   Kalman Filter Predictions for China

In Fig. 4.1 are presented 5 time series of active cases in China, considered as testing samples and the corresponding prediction for each of them.

The model obtained an average R^2 score metric of -7.6301 and an average MAE of 716.76. For the MSE, since it is a quadratic measure, the average doesn't make really sense to consider, because the penalties for larger individual errors are emphasized. For China, considering the proposed model, the R^2 Score was negative for samples 3, 4, and 5. Negative R^2 Scores normally means that the approximation curve is worse than the mean. Analyzing the curves through visual inspection, it can be seen that the convergence of the prediction takes a significant number of samples and this error in the beginning reflects into all the evaluation criteria. Besides, the error metrics such as MAE are also considered in the analysis.

Excluding the negative results, the best R^2 Score performance was 0.8111, achieved for Sample 1, presented in Fig. 4.1a, while the worse result is 0.7104 for Sample 2.

According to the results, the lowest MAE Score was 109.6761 for Sample 5, while the highest MAE Score was 2,677.10 for Sample 1. The same for MSE, Sample 5 presented the lowest quadratic error, while Sample 1 MSE was the highest.

A careful interpretation of the evaluation metrics shall be performed, since the sample with the best (lowest) MAE, presented the lowest negative R^2 Score. On the other hand, for sample 1, the best R^2 Score also presented the worst (highest) MAE. This means that, for the classification, one single metric should not be considered alone. For an overall evaluation of the predictor, it is necessary to assess different criteria and multiple results. Besides, the achieved results for other countries were different from the predictions related to China and are presented in the following sections.

4.3.2 Kalman Filter Predictions for the United States

In Fig. 4.2, are presented 5 time series of active cases in the United States of America, considered as testing samples and the corresponding prediction for each of them.

The model obtained an average R^2 score metric of 0.9377 and an average MAE of 4,439.39. For the MSE, since it is a quadratic measure, the average doesn't make really sense to consider, because the penalties for larger individual errors are emphasized.

For the US, on the contrary of what was found for China, considering the proposed model, there was no negative R^2 Score. Analyzing the curves through visual inspection, it can be seen that the prediction model worked well even for some samples with oscillating behavior, such as Sample 3 and 5. The best R^2 Score performance was 0.9894, achieved for Sample 2, presented in Fig 4.2b, while the worse result is 0.8012 for Sample 5.

According to the results, the lowest MAE Score was 3.7551 for Sample 1, while the highest MAE Score was 14,663.03 for Sample 5. The same for MSE, Sample 1 presented the lowest quadratic error, while Sample 5 MSE was the highest.

4.3.3 Kalman Filter Predictions for Brazil

In Fig. 4.3, are presented 5 time series of active cases in Brazil, considered as testing samples and the corresponding prediction for each of them.

The model obtained an average R^2 score metric of -0.8853 and an average MAE of 2,601.68. For the MSE, since it is a quadratic measure, the average doesn't make really sense to consider, because the penalties for larger individual errors are emphasized.

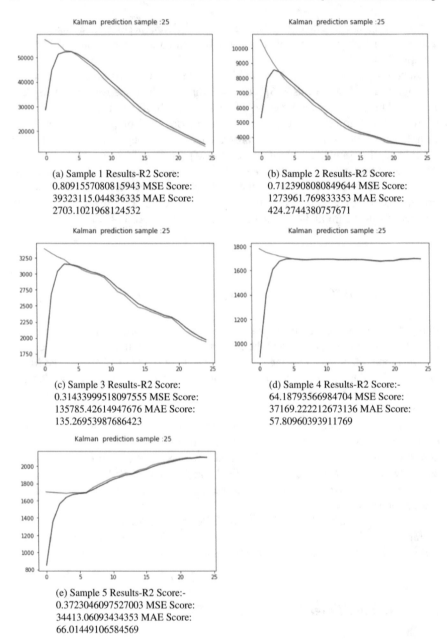

(a) Sample 1 Results-R2 Score:
0.8091557080815943 MSE Score:
39323115.044836335 MAE Score:
2703.1021968124532

(b) Sample 2 Results-R2 Score:
0.7123908080849644 MSE Score:
1273961.769833353 MAE Score:
424.2744380757671

(c) Sample 3 Results-R2 Score:
0.31433999518097555 MSE Score:
135785.42614947676 MAE Score:
135.26953987686423

(d) Sample 4 Results-R2 Score:-
64.18793566984704 MSE Score:
37169.222212673136 MAE Score:
57.80960393911769

(e) Sample 5 Results-R2 Score:-
0.3723046097527003 MSE Score:
34413.06093434353 MAE Score:
66.01449106584569

Fig. 4.1 Prediction Results of the Kalman Filter model for 5 samples of COVID-19 Active Cases time series from China. The blue lines are the original data, while the red lines are the prediction time series

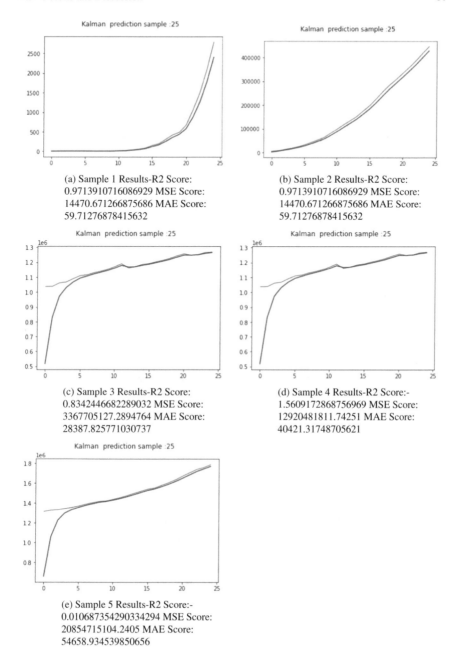

(a) Sample 1 Results-R2 Score: 0.9713910716086929 MSE Score: 14470.671266875686 MAE Score: 59.71276878415632

(b) Sample 2 Results-R2 Score: 0.9713910716086929 MSE Score: 14470.671266875686 MAE Score: 59.71276878415632

(c) Sample 3 Results-R2 Score: 0.8342446682289032 MSE Score: 3367705127.2894764 MAE Score: 28387.825771030737

(d) Sample 4 Results-R2 Score:- 1.5609172868756969 MSE Score: 12920481811.74251 MAE Score: 40421.31748705621

(e) Sample 5 Results-R2 Score:- 0.010687354290334294 MSE Score: 20854715104.2405 MAE Score: 54658.934539850656

Fig. 4.2 Prediction Results of the Kalman Filter model for 5 samples of COVID-19 Active Cases time series from the United States. The blue lines are the original data, while the red lines are the prediction time series

For Brazil, considering the proposed model, the R^2 Score was negative for sample 3 and brought the average to a negative value too. Negative R^2 Scores normally means that the approximation curve is worse than the mean. Analyzing the specific curve through visual inspection, it can be seen that the convergence of the prediction takes a significant number of samples and this error in the beginning reflects into all the evaluation criteria. Besides, the error metrics such as MAE are also considered in the analysis. Excluding the negative result, the best R^2 Score performance was 0.9866, achieved for Sample 1, presented in Fig. 4.3a, while the worse result is 0.4699 for Sample 4.

According to the results, the lowest MAE Score was 467.95 for Sample 1, while the highest MAE Score was 4, 245.91 for Sample 4. The same for MSE, Sample 1 presented the lowest quadratic error, while Sample 4 MSE was the highest. In this analysis, sample 4 presented the lowest R^2 Score and also the highest MAE result, indicating an example with poor prediction.

4.3.4   Kalman Filter Predictions for Italy

In Fig. 4.4, are presented 5 time series of active cases in Italy, considered as testing samples and the corresponding prediction for each of them.

The model obtained an average R^2 score metric of 0.2646 and an average MAE of 23, 702.11. For the MSE, since it is a quadratic measure, the average doesn't make really sense to consider, because the penalties for larger individual errors are emphasized.

For Italy, considering the proposed model, the R^2 Score was negative for samples 4 (this one very close to 0) and 5, what results in an average with a lower value. Negative R^2 Scores normally means that the approximation curve is worse than the mean. Analyzing the curves through visual inspection, it can be seen that the convergence of the prediction takes a significant number of samples and this error in the beginning reflects into all the evaluation criteria. Besides, the error metrics such as MAE are also considered in the analysis. Excluding the negative results, the best R^2 Score performance was 0.9921, achieved for Sample 2, presented in Fig. 4.4b, while the worse result was 0.9125 for Sample 3.

According to the results, the lowest MAE Score was 66.98 for Sample 1, while the highest MAE Score was 46, 623.21 for Sample 5. The same for MSE, Sample 1 presented the lowest quadratic error, while Sample 5 MSE was the highest. In this analysis, sample 5 presented a negative R^2 Score and also the highest MAE result, indicating an example with poor prediction.

4.3.5   Kalman Filter Predictions for Singapore

In Fig. 4.5, are presented 5 time series of active cases in Singapore, considered as testing samples and the corresponding prediction for each of them.

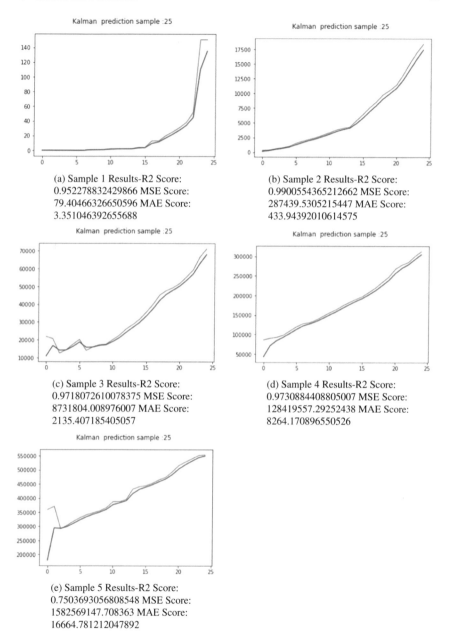

(a) Sample 1 Results-R2 Score: 0.952278832429866 MSE Score: 79.40466326650596 MAE Score: 3.351046392655688

(b) Sample 2 Results-R2 Score: 0.9900554365212662 MSE Score: 287439.5305215447 MAE Score: 433.94392010614575

(c) Sample 3 Results-R2 Score: 0.9718072610078375 MSE Score: 8731804.008976007 MAE Score: 2135.407185405057

(d) Sample 4 Results-R2 Score: 0.9730884408805007 MSE Score: 128419557.29252438 MAE Score: 8264.170896550526

(e) Sample 5 Results-R2 Score: 0.7503693056808548 MSE Score: 1582569147.708363 MAE Score: 16664.781212047892

Fig. 4.3 Prediction Results of the Kalman Filter model for 5 samples of COVID-19 Active Cases time series from Brazil. The blue lines are the original data, while the red lines are the prediction time series

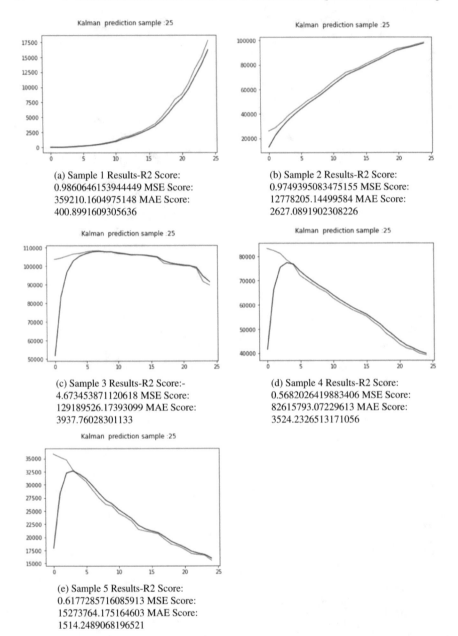

(a) Sample 1 Results-R2 Score:
0.9860646153944449 MSE Score:
359210.1604975148 MAE Score:
400.8991609305636

(b) Sample 2 Results-R2 Score:
0.9749395083475155 MSE Score:
12778205.14499584 MAE Score:
2627.0891902308226

(c) Sample 3 Results-R2 Score:-
4.673453871120618 MSE Score:
129189526.17393099 MAE Score:
3937.76028301133

(d) Sample 4 Results-R2 Score:
0.5682026419883406 MSE Score:
82615793.07229613 MAE Score:
3524.2326513171056

(e) Sample 5 Results-R2 Score:
0.6177285716085913 MSE Score:
15273764.175164603 MAE Score:
1514.2489068196521

Fig. 4.4 Prediction Results of the Kalman Filter model for 5 samples of COVID-19 Active Cases
time series from Italy. The blue lines are the original data, while the red lines are the prediction
time series

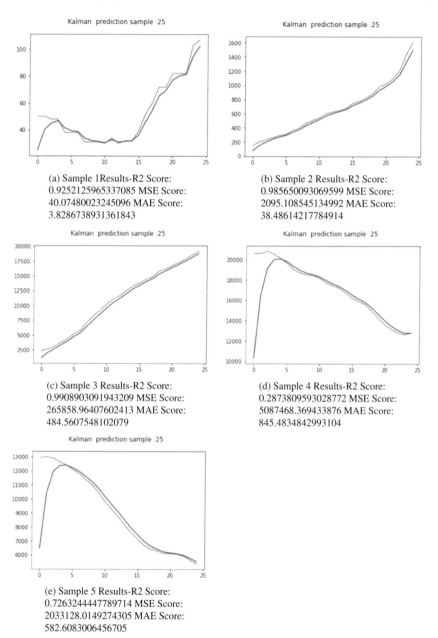

(a) Sample 1Results-R2 Score: 0.9252125965337085 MSE Score: 40.07480023245096 MAE Score: 3.8286738931361843

(b) Sample 2 Results-R2 Score: 0.985650093069599 MSE Score: 2095.108545134992 MAE Score: 38.48614217784914

(c) Sample 3 Results-R2 Score: 0.9908903091943209 MSE Score: 265858.96407602413 MAE Score: 484.5607548102079

(d) Sample 4 Results-R2 Score: 0.2873809593028772 MSE Score: 5087468.369433876 MAE Score: 845.4834842993104

(e) Sample 5 Results-R2 Score: 0.7263244447789714 MSE Score: 2033128.0149274305 MAE Score: 582.6083006456705

Fig. 4.5 Prediction Results of the Kalman Filter model for 5 samples of COVID-19 Active Cases time series from Singapore. The blue lines are the original data, while the red lines are the prediction time series

The model obtained an average R^2 score metric of 0.5296 and an average MAE of 364.62. For the MSE, since it is a quadratic measure, the average doesn't make really sense to consider, because the penalties for larger individual errors are emphasized.

For Singapore, considering the proposed model, the R^2 Score was negative for Sample 4 and this affected the average to a lower value too. Negative R^2 Scores normally means that the approximation curve is worse than the mean. Analyzing the specific curve through visual inspection, it can be seen that the convergence of the prediction takes a significant number of samples and this error in the beginning reflects into all the evaluation criteria. Besides, the error metrics such as MAE are also considered in the analysis. Excluding the negative result, the best R^2 Score performance was 0.9928, achieved for Sample 3, presented in Fig. 4.5c, while the worse result is 0.1903 for Sample 5.

According to the results, the lowest MAE Score was 3.99 for Sample 1, while the highest MAE Score was 758.43 for Sample 4. The same for MSE, Sample 1 presented the lowest quadratic error, while Sample 4 MSE was the highest. In this analysis, sample 4 presented a negative R^2 Score and also the highest MAE result, indicating an example with poor prediction.

4.3.6   Model Performance Comparison Between Countries

The last analysis that may be considered is the comparison of the best performance of the Kalman Filter Model between the five different countries considered in this work. The results are presented in Table 4.1.

The results of R^2 Score are presented in Table 4.2, not considering negative values. It is important to notice that, for this score, the highest the results, the better, since this is a ratio, i.e., a measure based on a percentage. On the other hand, the best and worst results of the MAE Score are in Table 4.3. This metric is a measure of a sum of differences for each sample, so the lowest the result, the better, with a lower accumulated error.

Table 4.1 Compilation of the best parameters for the Kalman Filter model with the average of R^2 Score and MAE Score for the 5 time series used for validation of each country

Country	Average R^2 score	Average MAE score
China	−7.63	716.76
US	0.93	4,439.39
Brazil	−0.88	2,601.68
Italy	0.26	23,702.11
Singapore	0.52	364.62

Table 4.2 Highest (best) and Lowest (worst) results of the R^2 Score, excluding negative values, for the Kalman Filter model considering the 5 time series used for validation of each country

Country	Highest R^2 Score	Lowest R^2 Score
China	0.8111	0.7104
US	0.9894	0.8012
Brazil	0.9866	0.4699
Italy	0.9921	0.9125
Singapore	0.9928	0.1903

Table 4.3 Lowest (best) and Highest (worst) results of the MAE Score for the Kalman Filter model considering the 5 time series used for validation of each country

Country	Lowest MAE	Highest MAE
China	109.67	2,677.10
US	3.75	14,663.03
Brazil	467.95	4,245.91
Italy	66.98	46,623.21
Singapore	3.99	758.43

4.4 Conclusion

According to the presented discussion, the use of Quadratic Kalman Filter as a predictor for the COVID-19 epidemiological data can be considered, with certain limitations being considered. The results of 5 different countries were presented with R^2 Score ranging from 0.19 when predicting the number of infected from Singapore to 0.99 for Italy and Singapore. On the other hand, the accumulated error (MAE) from the US was the lowest, 3.75, while for Italy it was the highest, 46, 623.21.

Higher R^2 Scores were obtained when the sample time series was smoothly increasing or decreasing and the initial prediction was closer to the initial sample data. The error metrics were higher when the prediction was performed for oscillating data series. Although the metrics didn't present satisfactory results, the visual inspection of the prediction plots could show that, after some nonfavorable initial conditions, the model could approximate the original data sample.

The application of the Quadratic Kalman Filter was discussed in this chapter. Another model based on Artificial Intelligence, considering the use of Long Short-Term Memory Neural Networks is presented in the following chapter allowing the reader to understand the impact of using an AI solution for the prediction of the COVID-19 epidemiological time series.

The next chapter will present the application of Artificial Intelligence (AI) techniques for the prediction task of the epidemiological time series of the COVID-19 pandemic. Two techniques are going to be considered. The first one is a neural network architecture called Long Short-Term Memory (LSTM), and the second considers the use of an automatic selector of different machine learning techniques, called AutoML, based on the best performance index for the proposed prediction.

References

1. B. Cazelles, N.P. Chau, Using the Kalman filter and dynamic models to assess the changing HIV/AIDS epidemic. Math. Biosci. **140**(2), 131–154 (1997). issn: 00255564. https://doi.org/10.1016/S0025-5564(96)00155-1
2. S. Haykin, *Kalman Filtering and Neural Networks*, vol. 47 (Wiley, 2004)
3. Kaggle, Kaggle: Your Home for Data Science (2020). https://www.kaggle.com/ (visited on 08/04/2020)
4. J. Mandel et al., Data driven computing by the morphing fast Fourier transform ensemble Kalman filter in epidemic spread simulations. Procedia Comput. Sci. **1**(1), 1221–1229 (2010). issn: 18770509. https://doi.org/10.1016/j.procs.2010.04.136. http://dx.doi.org/10.1016/j.procs.2010.04.136
5. Pandas, pandas - Python Data Analysis Library (2020). https://pandas.pydata.org/ (visited on 08/04/2020)
6. Python, Welcome to Python.org (2020). https://www.python.org/ (visited on 08/04/2020)
7. R.J.M. Singpurwalla, D. Nozer, Understanding the Kalman Filter (2005)
8. Statsmodel, Statsmodels - Python Package (2020). https://www.statsmodels.org/stable/index.html (visited on 08/04/2020)
9. Jeffrey K. Uhlmann, Simon J. Julier, A new extension of the kalman filter to nonlinear systems, in *Signal Processing, Sensor Fusion, and Target Recognition VI*, vol. 3068, pp. 182–194 (1997)
10. W. Yang, A. Karspeck, J. Shaman, Comparison of filtering methods for the modeling and retrospective forecasting of influenza epidemics. PLoS Comput. Biol. **10**(4) (2014). issn: 15537358. https://doi.org/10.1371/journal.pcbi.1003583

Chapter 5
Artificial Intelligence Prediction for the COVID-19 Data Based on LSTM Neural Networks and H2O AutoML

5.1 Introduction

The majority of the procedures considered in previous studies are more focused on strategies and neglect the elements intrinsic to the information. Several approaches rely on linear techniques and fail to catch nonlinear dynamics of the transmission rates which are characteristics of each disease. Statistical models like Autoregressive Moving Average (ARIMA), Moving Average (MA), Autoregressive (AR) systems are conventional methods used to forecasting real-time transmission prices. The use of R_0 is a statistical metric utilized as a key parameter to simulate disease. Referred to as reproduction number, R_0 forecasts the number of individuals can transmit the disease. If the value of R_0 of a disorder is 10, then the contaminated individual will spread the illness to 10 other people surrounding him [2]. Time-series forecast is the procedure of predicting upcoming trends/patterns of this specified dataset with capabilities that are temporal. A Time-Series (TS) data could be broken down into trend, seasonality, and error. A tendency in TS could be seen when a particular pattern repeats regular periods of time because of outside factors such as lockdown of state, compulsory social distancing, and quarantines [2].

Recurrent Long Short-Term Memory (LSTM) networks possess the capacity to deal with the weaknesses of time series forecasting methods by adapting nonlinearities. Generally speaking, Recurrent Neural Networks (RNN) are a kind of neural networks that aim to catch the dynamics of a time-series order. With this purpose, a state that could reveal information is retained by the network. In earlier times, since the RNN model's practice contains countless parameters, one significant challenge is to train the RNN effectively. Nonetheless, in the past few decades, together with the maturation of computing technology, network design, and computing, it became possible to train different versions of RNNs and for one of these, the models are structured on the Long Short-Term Memory (LSTM) model, while Bidirectional RNN (BRNN) presents a singular breakthrough in functionality and can complete many different tasks [2, 5].

© The Author(s), under exclusive license to Springer Nature Switzerland AG 2021
J. A. L. Marques et al., *Predictive Models for Decision Support
in the COVID-19 Crisis*, SpringerBriefs in Applied Sciences and Technology,
https://doi.org/10.1007/978-3-030-61913-8_5

This chapter focus on the application of LSTM RNN for the prediction of epidemiological data from five different countries during the COVID-19 pandemic. Besides that, a new tool called H2O Automatic Machine Learning (H2O AutoML) is also tested and discussed in the results section. This tool is presented in more detail in the following sections but as a brief presentation it aims to submit the time series under study to several different machine learning techniques and select the best results achieved by each of them. This significantly reduces the modeling and development time and allows the selection of the most suitable architecture for prediction and support the decision-making process more effectively.

5.2   Materials and Methods

5.2.1   Long Short-Term Memory Networks (LSTM)

Long Short-Term Memory networks are a sort of Recurrent Neural Networks. LSTMs are a suitable methodology for prediction tasks. The model creates future forecasts using several features existing in the datasets. The information flows with LSTMs with parts named as cell states. Since RNNs can save only a limited quantity of data, Long-Term Memory Storage cells (LSTM) are used alongside RNNs. LSTMs defeat the issues of vanishing gradient and exploding gradient [1]. Each block of LSTM performs at different time levels and passes its output to the subsequent block until the last LSTM piece produces the sequential output [2]. LSTM will initially pass a forgetting gate and pass to a sigmoid layer, which is also called the forgetting gate layer. Take the state h and the input x, and the output is a value between [0,1]. The decision on C_{t-1} means completely forgotten, 1 means completely reserved, which is [5]:

$$f_t = \sigma(w_f[h, x] + b_f)$$

$$G_t = \sigma(w_g[h, x] + b_g)$$

$$P_t = \sigma(w_p[h, x] + b_p)$$

Where

- f_t = function of input gate,
- G_t = function of forget gate,
- P_t = function of output gate,
- x = input to the current function at time step,
- H = result from the previous time step,
- W_f = coefficients of neurons at gate (x), and

- b_f = bias of neurons at gate (x).

Figure 5.1 present the LSTM model used in the experiments. Each LSTM model was training by 200 epochs and a Learning Rate on Plateau callback with a factor of 0.2 and patience set to 2.

5.2.2 H2O AutoML

H2O is an open-source, distributed machine learning platform built to scale to large datasets and APIs in Java, Python, R, and Scala. The tool provides H2O AutoML, a learning algorithm that piled ensembles within a purpose and overlooks the process of find candidate models. The consequence of the AutoML series is a rated list of best models for a dataset. Models in the leader-board could be rated by design performance metrics or version features like typical forecast rate or coaching time. H2O AutoML uses the combination of random grid search with stacked ensembles, as diversified models improve the ensemble method's accuracy [4].

5.2.3 Proposed Methodology for the Data Analysis

A general description about the considered methodology for Chaps. 3–5 of the book is presented in Chap. 1, Sect. 1.3. In that section are presented the source of COVID-19 related data; the justification of considering five countries on the analysis; the data handling approach (k-fold cross-validation strategy) and, finally, the evaluation criteria, considering R2 Score, Mean Square Error (MSE), and Mean Absolute Error (MAE). Essential references for each topic are also presented there.

The development environment was structured as a Python notebook (Python 2020), hosted by Kaggle (Kaggle 2020), including data analysis and manipulation packages, such as NumPy and Pandas (Pandas 2020) and Keras [3].

The results are presented in the next section, organized in the following sequence of countries: China, USA, Brazil, Italy, and Singapore.

5.3 Results and Discussion

In this chapter, the results are presented according to the following steps. For each country, a figure with five-time series selected during the cross-validation phase is presented with their corresponding prediction. The three evaluation criteria, R^2, MSE, and MAE, are discussed and compared. The LSTM model was applied with the input of 25 steps and a single output. The outputs for every 25 steps in each sample are presented in the next sections (Fig. 5.2).

Fig. 5.1 LSTM model used
for the experiments

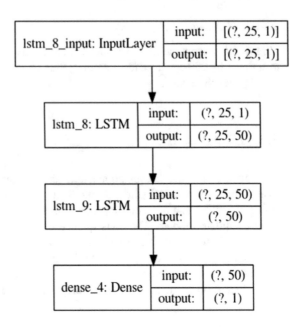

5.3.1   LSTM Predictions for China

In Fig. 5.3, are presented a time series of active cases in China, considered as testing
samples and the corresponding prediction for each of them.

The model obtained an average R^2 score metric of 0.92 and an average MAE of
12, 161.6. For the MSE, since it is a quadratic measure, the average doesn't make
really sense to consider, because the penalties for larger individual errors are emphasized. For China, considering the proposed model. Analyzing the curves through
visual inspection, it can be seen that the convergence of the prediction takes a significant number of samples. Besides, the error metrics such as MAE are also considered
in the analysis.

5.3.2   H2O AutoML Results for China

Table 5.1 presents the results for the H2O AutoML applied to China dataset. The
model ID shows the best models chosen.

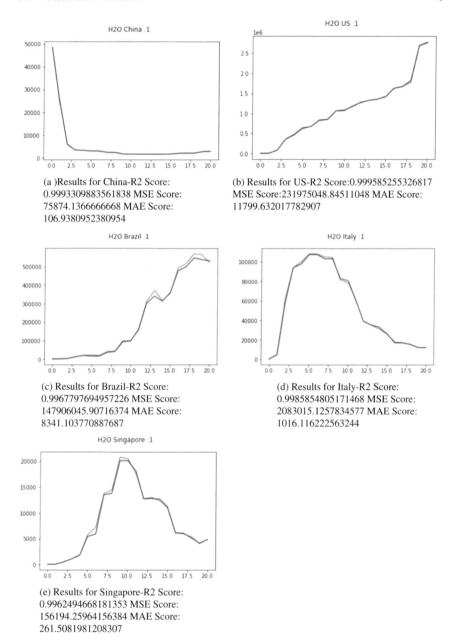

(a)Results for China-R2 Score: 0.9993309883561838 MSE Score: 75874.1366666668 MAE Score: 106.9380952380954

(b) Results for US-R2 Score:0.999585255326817 MSE Score:231975048.84511048 MAE Score: 11799.632017782907

(c) Results for Brazil-R2 Score: 0.9967797694957226 MSE Score: 147906045.90716374 MAE Score: 8341.103770887687

(d) Results for Italy-R2 Score: 0.9985854805171468 MSE Score: 2083015.1257834577 MAE Score: 1016.116222563244

(e) Results for Singapore-R2 Score: 0.9962494668181353 MSE Score: 156194.25964156384 MAE Score: 261.5081981208307

Fig. 5.2 Prediction Results of the H2O AutoML model for 1 sample of COVID-19 Active Cases time series from China, US, Brazil, Italy, and Singapore. The blue lines are the original data, while the red lines are the prediction time series

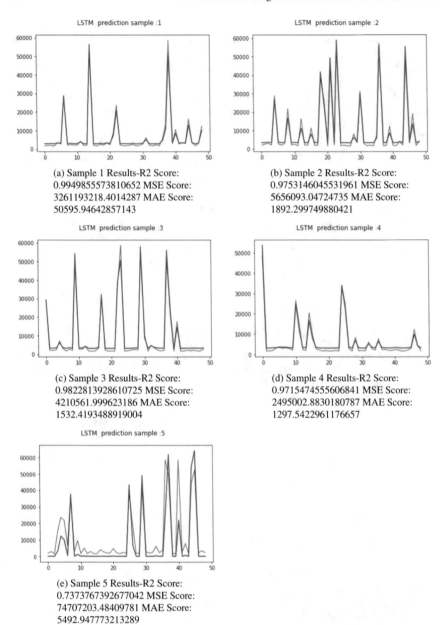

(a) Sample 1 Results-R2 Score: 0.9949855573810652 MSE Score: 3261193218.4014287 MAE Score: 50595.94642857143

(b) Sample 2 Results-R2 Score: 0.9753146045531961 MSE Score: 5656093.04724735 MAE Score: 1892.299749880421

(c) Sample 3 Results-R2 Score: 0.9822813928610725 MSE Score: 4210561.999623186 MAE Score: 1532.4193488919004

(d) Sample 4 Results-R2 Score: 0.9715474555606841 MSE Score: 2495002.8830180787 MAE Score: 1297.5422961176657

(e) Sample 5 Results-R2 Score: 0.7373767392677042 MSE Score: 74707203.48409781 MAE Score: 5492.947773213289

Fig. 5.3 Prediction Results of the LSTM model for 5 samples of COVID-19 Active Cases time series from China. The blue lines are the original data, while the red lines are the prediction time series

Table 5.1 H2O AutoML results for China prediction

Model	Mean residual deviance	RMSE	MSE	MAE	RMSLE
XRT_1_AutoML_20200817_030702	75874.1	275.453	75874.1	106.938	0.0228265
DRF_1_AutoML_20200817_030702	137789	371.2	137789	135.483	0.0277021
GBM_grid__1_AutoML_20200817_030702_model_2	158954	398.69	158954	257.486	0.0678476
StackedEnsemble_BestOfFamily_AutoML_20200817_030702	168276	410.214	168276	161.928	0.0296967
StackedEnsemble_AllModels_AutoML_20200817_030702	360471	600.393	360471	481.159	0.166096
XGBoost_grid__1_AutoML_20200817_030702_model_3	446086	667.897	446086	347.258	0.221459
XGBoost_3_AutoML_20200817_030702	527290	726.147	527290	350.009	0.104589
GBM_4_AutoML_20200817_030702	709494	842.315	709494	442.611	0.104531
GBM_grid__1_AutoML_20200817_030702_model_1	931395	965.088	931395	391.759	0.0864024
XGBoost_grid__1_AutoML_20200817_030702_model_2	940096	969.586	940096	350.033	0.042638
DeepLearning_grid__3_AutoML_20200817_030702_model_1	1.47281e+06	1213.59	1.47281e+06	931.235	0.354891
XGBoost_2_AutoML_20200817_030702	1.48597e+06	1219	1.48597e+06	597.542	0.155796
DeepLearning_1_AutoML_20200817_030702	1.58251e+06	1257.98	1.58251e+06	795.047	0.220298
GBM_2_AutoML_20200817_030702	1.90232e+06	1379.25	1.90232e+06	970.354	0.269377
XGBoost_1_AutoML_20200817_030702	1.91741e+06	1384.71	1.91741e+06	506.634	0.0712718
GBM_3_AutoML_20200817_030702	1.94218e+06	1393.62	1.94218e+06	866.098	0.224971
XGBoost_grid__1_AutoML_20200817_030702_model_1	2.24767e+06	1499.22	2.24767e+06	523.246	0.0644726
GBM_1_AutoML_20200817_030702	2.2963e+06	1515.35	2.2963e+06	1194.49	0.348035
DeepLearning_grid__2_AutoML_20200817_030702_model_1	2.32746e+06	1525.6	2.32746e+06	1309.19	0.680922
DeepLearning_grid__1_AutoML_20200817_030702_model_1	6.06969e+06	2463.67	6.06969e+06	1068.5	0.183813
DeepLearning_grid__1_AutoML_20200817_030702_model_2	2.72146e+07	5216.76	2.72146e+07	5114.81	nan
GLM_1_AutoML_20200817_030702	1.21174e+08	11007.9	1.21174e+08	8059	1.33252

5.3.3   LSTM Predictions for the United States

In Fig. 5.4, are presented 5 time series of active cases in the United States of America, considered as testing samples and the corresponding prediction for each of them.

The model obtained an average R^2 score metric of 0.79 and an average MAE of 13, 251.4. For the MSE, since it is a quadratic measure, the average doesn't make really sense to consider, because the penalties for larger individual errors are emphasized.

Analyzing the curves through visual inspection, it can be seen that the prediction model worked except for Sample 4. The best R^2 Score performance was 0.994, achieved for Sample 1, presented in Fig. 5.2a, while the worse result is 0.33 for Sample 4.

According to the results, the lowest MAE Score was 1, 338 for Sample 3, while the highest MAE Score was 50, 595 for Sample 1.

5.3.4   H2O AutoML Results for US

Table 5.2 presents the results for the H2O AutoML applied to US dataset. The model ID shows the best models chosen.

5.3.5   LSTM Predictions for Brazil

In Fig. 5.5, are presented 5 time series of active cases in Brazil, considered as testing samples and the corresponding prediction for each of them.

The model obtained an average R^2 score metric of 0.93 and an average MAE of 40, 604.4. For the MSE, since it is a quadratic measure, the average doesn't make really sense to consider, because the penalties for larger individual errors are emphasized.

For Brazil, considering the proposed model and analyzing the specific curve through visual inspection, it can be seen that the convergence of the prediction takes a significant number of samples and this error in the beginning reflects into all the evaluation criteria. Besides, the error metrics such as MAE are also considered in the analysis. The best R^2 Score performance was 0.9939, achieved for Sample 3, presented in Fig. 5.3c, while the worse result is 0.9232 for Sample 2.

According to the results, the lowest MAE Score was 10, 437.99 for Sample 3, while the highest MAE result was 50, 401.91 for Sample 5. In this analysis, Sample 2 presented the lowest R^2 Score and also the highest MAE.

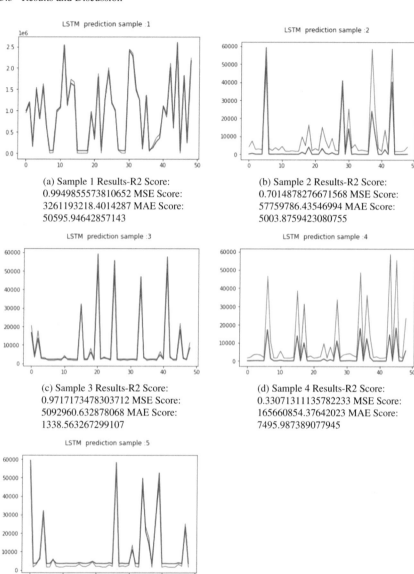

(a) Sample 1 Results-R2 Score:
0.9949855573810652 MSE Score:
3261193218.4014287 MAE Score:
50595.94642857143

(b) Sample 2 Results-R2 Score:
0.7014878276671568 MSE Score:
57759786.43546994 MAE Score:
5003.8759423080755

(c) Sample 3 Results-R2 Score:
0.9717173478303712 MSE Score:
5092960.632878068 MAE Score:
1338.563267299107

(d) Sample 4 Results-R2 Score:
0.33071311135782233 MSE Score:
165660854.37642023 MAE Score:
7495.987389077945

(e) Sample 5 Results-R2 Score:
0.9786482452656845 MSE Score:
4420464.358780392 MAE Score:
1826.1186971859056

Fig. 5.4 Prediction Results of the LSTM model for 5 samples of COVID-19 Active Cases time series from the United States. The blue lines are the original data, while the red lines are the prediction time series

Table 5.2 H2O AutoML results for US prediction

Model	Mean residual deviance	RMSE	MSE	MAE	RMSLE
StackedEnsemble_BestOfFamily_AutoML_20200817_032018	2.31975e+08	15230.7	2.31975e+08	11799.6	1.40355
DRF_1_AutoML_20200817_032018	3.55707e+08	18860.2	3.55707e+08	13622.4	0.226362
XRT_1_AutoML_20200817_032018	6.60015e+08	25690.8	6.60015e+08	18583.5	0.182281
GBM_grid__1_AutoML_20200817_032018_model_2	7.61682e+08	27598.6	7.61682e+08	18867.2	0.628974
XGBoost_grid__1_AutoML_20200817_032018_model_1	8.2726e+08	28762.1	8.2726e+08	23561.6	0.579766
XGBoost_grid__1_AutoML_20200817_032018_model_2	1.15224e+09	33944.6	1.15224e+09	28047.3	nan
XGBoost_grid__1_AutoML_20200817_032018_model_3	1.68925e+09	41100.4	1.68925e+09	30419.2	1.22456
XGBoost_3_AutoML_20200817_032018	4.2481e+09	65177.5	4.2481e+09	49096.6	0.895225
GBM_grid__1_AutoML_20200817_032018_model_1	4.27328e+09	65370.3	4.27328e+09	38237.9	nan
XGBoost_1_AutoML_20200817_032018	5.11981e+09	71552.8	5.11981e+09	50076.1	0.960027
GBM_3_AutoML_20200817_032018	6.05067e+09	77786	6.05067e+09	45295.1	1.81981
GBM_2_AutoML_20200817_032018	7.63173e+09	87359.8	7.63173e+09	47985.1	1.96917
DeepLearning_1_AutoML_20200817_032018	8.25332e+09	90847.8	8.25332e+09	57017.3	nan
XGBoost_2_AutoML_20200817_032018	8.7999e+09	93807.8	8.7999e+09	53561.9	1.17093
StackedEnsemble_AllModels_AutoML_20200817_032018	9.30835e+09	96479.8	9.30835e+09	62868.7	2.38154
GBM_4_AutoML_20200817_032018	1.03687e+10	101827	1.03687e+10	55632.1	2.12958
GBM_1_AutoML_20200817_032018	1.49039e+10	122082	1.49039e+10	93351	2.48399
DeepLearning_grid__1_AutoML_20200817_032018_model_2	5.8085e+10	241008	5.8085e+10	222085	nan
GBM_grid__1_AutoML_20200817_032018_model_4	7.332e+10	270777	7.332e+10	144059	2.43224
DeepLearning_grid__1_AutoML_20200817_032018_model_1	8.1683e+10	285802	8.1683e+10	270933	nan
DeepLearning_grid__2_AutoML_20200817_032018_model_1	9.83231e+10	313565	9.83231e+10	247422	nan
DeepLearning_grid__3_AutoML_20200817_032018_model_1	1.01252e+11	318200	1.01252e+11	247869	2.55519
GBM_grid__1_AutoML_20200817_032018_model_5	5.12041e+11	715570	5.12041e+11	530956	3.10216
GLM_1_AutoML_20200817_032018	5.88004e+11	766814	5.88004e+11	598850	3.13743

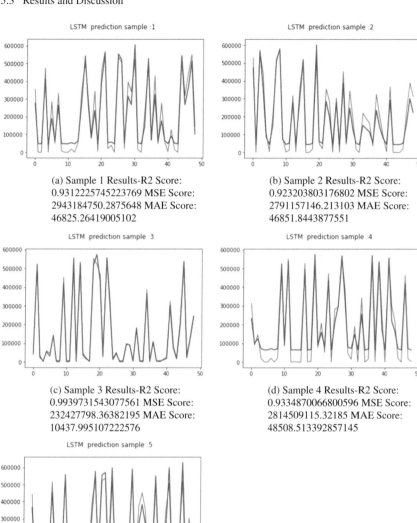

(a) Sample 1 Results-R2 Score:
0.9312225745223769 MSE Score:
2943184750.2875648 MAE Score:
46825.26419005102

(b) Sample 2 Results-R2 Score:
0.923203803176802 MSE Score:
2791157146.213103 MAE Score:
46851.8443877551

(c) Sample 3 Results-R2 Score:
0.9939731543077561 MSE Score:
232427798.36382195 MAE Score:
10437.995107222576

(d) Sample 4 Results-R2 Score:
0.9334870066800596 MSE Score:
2814509115.32185 MAE Score:
48508.513392857145

(e) Sample 5 Results-R2 Score:0.92919206580709
MSE Score:3192973911.363407 MAE Score:
50401.81991390306

Fig. 5.5 Prediction Results of the LSTM model for 5 samples of COVID-19 Active Cases time series from Brazil. The blue lines are the original data, while the red lines are the prediction time series

5.3.6 H2O AutoML Results for Brazil

Table 5.3 presents the results for the H2O AutoML applied to Brazil dataset. The model ID shows the best models chosen.

5.3.7 LSTM Predictions for Italy

In Fig. 5.6 are presented 5 time series of active cases in Italy, considered as testing samples and the corresponding prediction for each of them.

The model obtained an average R^2 score metric of 0.95 and an average MAE of 6, 336.8. For the MSE, since it is a quadratic measure, the average doesn't make really sense to consider, because the penalties for larger individual errors are emphasized.

For Italy, considering the proposed model. The best R^2 Score performance was 0.9745, achieved for Sample 3, presented in Fig. 5.4c, while the worse result was 0.9399 for Sample 2.

According to the results, the lowest MAE Score was 4, 493.21 for Sample 1, while the highest MAE Score was 6, 896.48 for Sample 2.

5.3.8 H2O AutoML Results for Italy

Table 5.4 presents the results for the H2O AutoML applied to Italy dataset. The model ID shows the best models chosen.

5.3.9 LSTM Predictions for Singapore

In Fig. 5.7, are presented 5 time series of active cases in Singapore, considered as testing samples and the corresponding prediction for each of them.

The model obtained an average R^2 score metric of 0.89 and an average MAE of 1, 622.2. For the MSE, since it is a quadratic measure, the average doesn't make really sense to consider, because the penalties for larger individual errors are emphasized.

Analyzing the specific curve through visual inspection, it can be seen that the convergence of the prediction takes a significant number of samples and this error in the beginning reflects into all the evaluation criteria. Besides, the error metrics such as MAE are also considered in the analysis. The best R^2 Score performance was 0.9347, achieved for Sample 2, presented in Fig. 5.5b, while the worse result is 0.8564 for Sample 1.

According to the results, the lowest MAE Score was 1277.70 for Sample 3, while the highest MAE Score was 2, 239.12 for Sample 1.

Table 5.3 H2O AutoML results for Brazil prediction

Model	Mean residual deviance	RMSE	MSE	MAE	RMSLE
StackedEnsemble_BestOfFamily_AutoML_20200817_033205	1.47906e+08	12161.7	1.47906e+08	8341.1	1.91057
XGBoost_grid__1_AutoML_20200817_033205_model_2	1.60829e+08	12681.9	1.60829e+08	8472.79	1.19059
XGBoost_grid__1_AutoML_20200817_033205_model_1	1.88194e+08	13718.4	1.88194e+08	8794.79	0.477489
XGBoost_grid__1_AutoML_20200817_033205_model_3	2.1877e+08	14790.9	2.1877e+08	9455.89	nan
GBM_grid__1_AutoML_20200817_033205_model_1	2.21982e+08	14899.1	2.21982e+08	10059.9	1.74582
DRF_1_AutoML_20200817_033205	2.39999e+08	15491.9	2.39999e+08	10249.1	0.359038
GBM_grid__1_AutoML_20200817_033205_model_2	3.12922e+08	17689.6	3.12922e+08	10971.7	1.1006
XGBoost_2_AutoML_20200817_033205	3.21528e+08	17931.2	3.21528e+08	12666.8	0.468404
XRT_1_AutoML_20200817_033205	3.70979e+08	19260.8	3.70979e+08	11948.7	0.133646
GBM_2_AutoML_20200817_033205	4.25397e+08	20625.2	4.25397e+08	14524.5	2.32287
GBM_grid__1_AutoML_20200817_033205_model_4	8.93975e+08	29899.4	8.93975e+08	19576.8	2.3569
GBM_3_AutoML_20200817_033205	1.07385e+09	32769.7	1.07385e+09	26647.3	2.74893
DeepLearning_grid__1_AutoML_20200817_033205_model_2	1.10781e+09	33283.8	1.10781e+09	27771.6	2.82175
XGBoost_3_AutoML_20200817_033205	1.11163e+09	33341.1	1.11163e+09	22708.6	1.16229
DeepLearning_grid__1_AutoML_20200817_033205_model_1	1.22473e+09	34996.2	1.22473e+09	31480.8	nan
GBM_4_AutoML_20200817_033205	1.31191e+09	36220.3	1.31191e+09	29617.6	2.79132
DeepLearning_1_AutoML_20200817_033205	1.34262e+09	36641.7	1.34262e+09	30949.3	3.0945
GBM_1_AutoML_20200817_033205	1.38445e+09	37208.2	1.38445e+09	31091.6	2.82896
XGBoost_1_AutoML_20200817_033205	1.53471e+09	39175.4	1.53471e+09	26662.2	0.200685
StackedEnsemble_AllModels_AutoML_20200817_033205	1.5516e+09	39390.4	1.5516e+09	32817.6	2.9552
DeepLearning_grid__2_AutoML_20200817_033205_model_1	4.02308e+09	63427.7	4.02308e+09	57918.7	nan
DeepLearning_grid__3_AutoML_20200817_033205_model_1	6.28681e+09	79289.4	6.28681e+09	70362.9	nan
GLM_1_AutoML_20200817_033205	4.6863e+10	216479	4.6863e+10	194287	3.7287

LSTM prediction sample :1

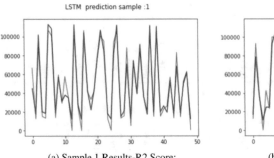

LSTM prediction sample :2

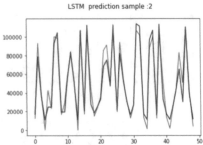

(a) Sample 1 Results-R2 Score:
0.950921468655632 MSE Score:
66062425.421406165 MAE Score:
6441.362065529337

(b) Sample 2 Results-R2 Score:
0.9399171739431909 MSE Score:
82041161.08867007 MAE Score:
6896.480747767857

LSTM prediction sample :3

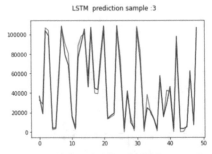

LSTM prediction sample :4

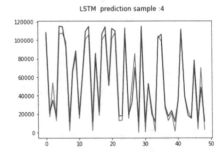

(c) Sample 3 Results-R2 Score:
0.9745482932572712 MSE Score:
39320332.33615532 MAE Score:
4943.218695192921

(d) Sample 4 Results-R2 Score:
0.9592610999052243 MSE Score:
67922031.25936984 MAE Score:
6702.508609693878

LSTM prediction sample :5

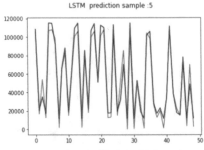

(e) Sample 5 Results-R2 Score:
0.9592610999052243 MSE Score:
67922031.25936984 MAE Score:
6702.508609693878

Fig. 5.6 Prediction Results of the LSTM model for 5 samples of COVID-19 Active Cases time
series from Italy. The blue lines are the original data, while the red lines are the prediction time
series

Table 5.4 H2O AutoML results for Italy prediction

Model	Mean residual deviance	RMSE	MSE	MAE	RMSLE
XGBoost_grid__1_AutoML_20200817_024752_model_2	2.08302e+06	1443.27	2.08302e+06	1016.12	0.0682207
StackedEnsemble_BestOfFamily_AutoML_20200817_024752	2.16323e+06	1470.79	2.16323e+06	1085.62	0.260525
XRT_1_AutoML_20200817_024752	2.23062e+06	1493.53	2.23062e+06	995.233	0.0504151
XGBoost_grid__1_AutoML_20200817_024752_model_1	2.7784e+06	1666.85	2.7784e+06	1274.61	0.0724224
DRF_1_AutoML_20200817_024752	3.95503e+06	1988.73	3.95503e+06	1261.96	0.0749394
XGBoost_2_AutoML_20200817_024752	4.07244e+06	2018.03	4.07244e+06	1652.68	0.293454
GBM_grid__1_AutoML_20200817_024752_model_1	6.63753e+06	2576.34	6.63753e+06	1898.69	0.369809
DeepLearning_grid__1_AutoML_20200817_024752_model_1	1.31883e+07	3631.57	1.31883e+07	2611.1	0.500503
StackedEnsemble_AllModels_AutoML_20200817_024752	2.74578e+07	5240.02	2.74578e+07	4637.82	0.697103
XGBoost_3_AutoML_20200817_024752	3.93392e+07	6272.1	3.93392e+07	4296.24	0.121593
GBM_3_AutoML_20200817_024752	1.10008e+08	10488.5	1.10008e+08	9414.4	0.844144
XGBoost_1_AutoML_20200817_024752	1.12001e+08	10583.1	1.12001e+08	7540	0.184732
GBM_2_AutoML_20200817_024752	1.31685e+08	11475.4	1.31685e+08	10283	0.867215
GBM_4_AutoML_20200817_024752	1.31747e+08	11478.1	1.31747e+08	10091.3	0.873631
GBM_1_AutoML_20200817_024752	2.47549e+08	15733.7	2.47549e+08	13907.6	0.985011
DeepLearning_1_AutoML_20200817_024752	9.90674e+08	31475	9.90674e+08	23879	0.991632
GLM_1_AutoML_20200817_024752	1.5497e+09	39366.2	1.5497e+09	34944	1.39688

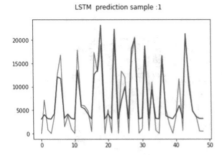

(a) Sample 1 Results-R2 Score:
0.8563405929365833 MSE Score:
7118576.688752386 MAE Score:
2239.1218610491073

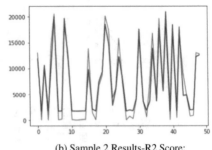

(b) Sample 2 Results-R2 Score:
0.9347138072941987 MSE Score:
3308472.9724531556 MAE Score:
1495.876051299426

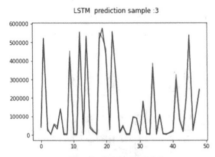

(c) Sample 3 Results-R2 Score:
0.9341933946241066 MSE Score:
2628372.5250954963 MAE Score:
1277.7034588249362

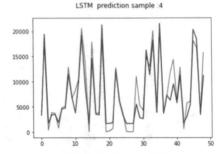

(d) Sample 4 Results-R2 Score:
0.884888941805073 MSE Score:
4825372.968090201 MAE Score:
1622.007476183833

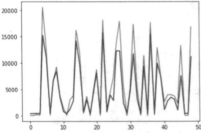

(e) Sample 5 Results-R2 Score:
0.8727746302089165 MSE Score:
4961388.040170446 MAE Score:
1478.515156648597

Fig. 5.7 Prediction Results of the LSTM model for 5 samples of COVID-19 Active Cases time series from Singapore. The blue lines are the original data, while the red lines are the prediction time series

Table 5.5 H2O AutoML results for Singapore prediction

Model	Mean residual deviance	RMSE	MSE	MAE	RMSLE
GBM_grid__1_AutoML_20200817_034229_model_1	156194	395.214	156194	261.508	0.0987321
StackedEnsemble_BestOfFamily_AutoML_20200817_034229	170151	412.493	170151	277.824	0.243919
GBM_grid__1_AutoML_20200817_034229_model_2	211226	459.594	211226	239.067	0.0923956
XGBoost_grid__1_AutoML_20200817_034229_model_3	230748	480.362	230748	267.421	0.152684
XGBoost_3_AutoML_20200817_034229	242944	492.894	242944	328.484	0.132666
XGBoost_2_AutoML_20200817_034229	265246	515.02	265246	345.613	0.211266
XGBoost_grid__1_AutoML_20200817_034229_model_2	314902	561.161	314902	355.352	0.10066
XGBoost_grid__1_AutoML_20200817_034229_model_1	342239	585.012	342239	383.742	0.313678
DRF_1_AutoML_20200817_034229	379171	615.768	379171	407.883	0.10363
XRT_1_AutoML_20200817_034229	467604	683.816	467604	369.062	0.122325
GBM_2_AutoML_20200817_034229	685037	827.67	685037	567.865	0.703202
XGBoost_1_AutoML_20200817_034229	731126	855.059	731126	602.025	0.152325
GBM_3_AutoML_20200817_034229	829272	910.644	829272	667.108	0.767878
StackedEnsemble_AllModels_AutoML_20200817_034229	883738	940.074	883738	805.728	1.01494
GBM_4_AutoML_20200817_034229	937245	968.114	937245	720.013	0.829386
DeepLearning_1_AutoML_20200817_034229	1.26952e+06	1126.73	1.26952e+06	732.946	0.239888
GBM_1_AutoML_20200817_034229	1.56955e+06	1252.82	1.56955e+06	1042.24	1.04457
DeepLearning_grid__1_AutoML_20200817_034229_model_1	4.66006e+06	2158.72	4.66006e+06	1806.32	1.03027
DeepLearning_grid__2_AutoML_20200817_034229_model_1	8.30888e+06	2882.51	8.30888e+06	2315.17	nan
DeepLearning_grid__1_AutoML_20200817_034229_model_2	8.64597e+06	2940.4	8.64597e+06	2675	nan
GLM_1_AutoML_20200817_034229	4.41893e+07	6647.5	4.41893e+07	5475.87	1.79249

5.3.10   H2O AutoML Results for Singapore

Table 5.5 presents the results for the H2O AutoML applied to Singapore dataset. The model ID shows the best models chosen.

5.4   Conclusion

The use of the LSTM model as a predictor for the COVID-19 epidemiological data can be considered. The results of 5 different countries were presented with a R^2 Average Score ranging from 0.89 when predicting the number of infected from Singapore to 0.95 for Italy. On the other hand, the accumulated error (MAE) from Singapore was the lowest, $1,622.72$, while for Brazil it was the highest, $40,604.44$.

Higher R^2 Scores were obtained when the sample time series was smoothly increasing or decreasing, and the initial prediction was closer to the initial sample data. The error metrics were higher when the prediction was performed for oscillating data series. Although the metrics did not present satisfactory results, the visual inspection of the prediction plots could show that, after some nonfavorable initial conditions, the model could approximate the original data sample.

The results obtained by the AutoML framework achieved higher R^2 Scores and lower MAE and MSE when compared with LSTM and also with other techniques proposed in the book, such as ARIMA and Kalman predictor, achieving $R^2 > 0.99$ for all of the 5 countries considered in this book. Other error metrics were also presented in case the reader is interested in comparing it with any other technique. This superior performance allows us to conclude that the application of a machine learning algorithm selector might be a promising candidate for a good predictor for epidemic time series.

The application of Artificial Intelligence tools for the prediction of the COVID-19 epidemiological curves was discussed in this chapter. A case study about a geographic predictor based on unsupervised methods for clustering and Kalman Filter for prediction is presented in the following chapter.

References

1. P. Arora, H. Kumar, B. Ketan Panigrahi, Prediction and analysis of COVID-19 positive cases using deep learning models: a descriptive case study of India. Chaos, Solitons Fractals **139** (2020). issn: 09600779. https://doi.org/10.1016/j.chaos.2020.110017
2. V.K.R. Chimmula, L. Zhang, Time series forecasting of COVID-19 transmission in Canada using LSTM networks. Chaos Solitons Fractals **135** (2020). issn: 09600779. https://doi.org/10.1016/j.chaos.2020.109864
3. Keras: the Python deep learning API (n.d.) (2020). https://keras.io/. (Accessed on 08/16/2020)
4. E. Ledell, S. Poirier, H2O AutoML: Scalable Automatic Machine Learning (2020) https://scinet.usda.gov/user/geospatial/

5. A. Zubair, Muhammad et al., Performance Evaluation Of Supervised Machine Learning Techniques For Efficient Detection Of Emotions From Online Content Design Optimization of Electric Machines View project Business Modeling View project Performance Evaluation Of Supervised Machine Learning Techniques For Efficient Detection Of Emotions From Online Content (2020). https://doi.org/10.32604/cmc.2020.xxxxx. www.techscience.com/journal/cmc

Chapter 6
Predicting the Geographic Spread of the COVID-19 Pandemic: A Case Study from Brazil

6.1 Introduction

The support provided by geographic data and the corresponding processing tools can play an essential role in several decision-making situations, especially during the current pandemic outbreak of the COVID-19. Several approaches considering geographic data are present in the literature. One report established the association between domestic train transportation and novel coronavirus (2019-nCoV) outbreak in China [9]. From another perspective, the effect of infection containment measures adopted by the Italian Government influenced geographically the spread and dynamics of the COVID-19 disease in the country [3].

From the managerial point of view, aspects such as preparedness, response, recovery, and mitigation must be considered in several dimensions, not only focusing on controlling or reducing the infection rate, but also when considering logistics aspects in public health such as number of ICU beds, designation of health professionals, allocation of consumables and critical equipment, procurement decision, among many others [5].

A comprehensive approach to several challenges during the struggle against the COVID-19 pandemic using Geographical Information Systems (GIS) and spatial big data is presented in [10]. According to the authors, the technology already has played an important role in identifying the spatial transmission of the epidemic, in spatial prevention and control of the epidemic, in spatial allocation of resources, and in spatial detection of social sentiment, among other things. Based on that analysis, ten challenges are presented and are relevant to be considered: (1) rapid construction of a big data information system for the epidemic; (2) rapid problem-oriented big data acquisition and integration; (3) convenient multi-scale dynamic mapping for epidemics; (4) comparison between spatial tracking and the spatiotemporal trajectory of big data; (5) spatiotemporal prediction of the transmission speed and scale of the epidemic; (6) spatial segmentation of the epidemic risk and prevention level; (7) spatial dynamic balancing of supply and demand for medical resources; (8) assessment

© The Author(s), under exclusive license to Springer Nature Switzerland AG 2021
J. A. L. Marques et al., *Predictive Models for Decision Support in the COVID-19 Crisis*, SpringerBriefs in Applied Sciences and Technology,
https://doi.org/10.1007/978-3-030-61913-8_6

of the supply of materials and transportation risk; (9) rapid estimation of the population flow and distribution; and (10) monitoring the spatial spread of social sentiment and detection.

Specifically, on the issue of data collection, one significant concern is about data quality and reliability [4]. For example, when collecting apparently simple patient data, such as the home address, which will be the most important information for the epidemic tracking and control measures, different sources of errors or imprecision are historically present. If there is a previous patient registration system available in the public health system, this makes the process easier. Nevertheless, in several countries and specifically in Brazil, which is the reality covered in this case study, this information is usually not integrated and not available to a large number of healthcare units. Besides, the infrastructure of healthcare units is poor.

For a reliable data collection, first, it is necessary to trust the patient when providing the information. Depending on the education level and stress conditions, this information may be provided incomplete or superficially. Second, the infrastructure of the healthcare center/unit is not standardized and while in some units a computerized registration can be performed, in others the registration may be performed on a manual form and follow the attendant limitations.

On the other hand, other strategies can be considered for precise geographic data collection. For example, to count on the ubiquitous use of smartphones with GPS and use the device to capture the patient address, ambulance location, or even tracking critical equipment.

The main objective of this chapter is to present a series of results following a case study format about the development of a geographic predictor for the City of Fortaleza, State of Ceara, Brazil, based on an implementation of a clustering approach and a two-dimensional Kalman predictor.

6.2   Proposed Methodology for the Geographic Epidemic Predictor

In this section, the proposed solution is described, including the details about the considered database, data collection process in general and the techniques used for clustering and geographic prediction.

The proposed tool estimates the latitude and longitude of new cases of COVID-19 based on the suggestion obtained from a group of unsupervised models for data clustering, followed by the application of a two-dimensional Kalman filter for the prediction.

6.2.1 *Location and Database*

The considered database was obtained in the City of Fortaleza by the Secretariat of Health of the Government of the State of Ceara. Fortaleza is the capital of Ceara and it's a city with more than 2 million inhabitants and a healthcare infrastructure organized in 4 levels, starting with the first level of ambulatory centers, another level also ambulatory units but for more complex cases and management of allocation to the third level which is formed by general hospitals. Additionally, a fourth level consists of specialized hospitals for specific cases. During the pandemic, the whole system was adapted to attend the COVID-19 with diagnostic, clinical exams, medical imaging, and hospitalization in different levels, from observation to ICU beds with artificial ventilation support.

Because of its touristic vocation and several flight links with other countries on a daily basis, Fortaleza became one of the most significant outbreaks during the first four months of the COVID-19 pandemic spread in Brazil. The Secretariat of Health established a task force to support the decisions of the Government of the State and one relevant task was to generate prediction on several levels, including geography. The system presented here is the result of one of these efforts.

In Fig. 6.1 is presented the evolution of the COVID-19 pandemic in the City of Fortaleza in 3 stages. At first, in January still with no domestic transmission registered. April, already with a strong spread and significant peak in the number of new cases per day until the end of July 2020. The geographic representation is based on the heatmap plot of the accumulated number of confirmed cases of COVID-19. The more red the region, the highest number of cases and the green part represents the lowest number of cases.

The original database considered a status classification for each monitored patient, such as suspicious, negative, or positive. For the current analysis, only the positive/confirmed cases are considered, since the goal is to predict where new cases are going to occur. Future analysis could also consider the suspicious cases as a reference for clustering data since the number of tests is restricted and only patients with any

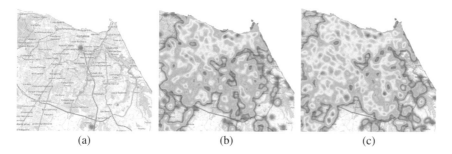

(a)	(b)	(c)

Fig. 6.1 Example of one heatmap of the accumulated number of confirmed cases of COVID-19 in the City of Fortaleza, Brazil. **a** January 2020 with no domestic cases. **b** At the end of April 2020, already with a significant spread in different regions of the City. **c** At the end of June 2020, after the implementation of a successful program of infection containment measures

symptom are submitted to clinical analysis. Considering the significant rate of false negative results, the suspicious cases can also indicate a trend for decision-makers of where the symptoms are more common.

6.2.2 Clustering and Prediction Techniques

In this approach, several algorithms were used to predict the next locations that affect the infection, using clustering methods we create of geographic groups and after use the Kalman filter to display the new points of infection.

The unsupervised models considered for data clustering were Agglomerative Clustering, DBSCAN, Mean Shift, and K-Means. They were executed in sequence, following a serial arrangement as presented in the flow diagram of Fig. 6.2

A brief introduction to each technique is presented in the following subsections. The development environment was a $Kaggle$ notebook using $Python$ programming language and the $SKLearn$ library was the main source for the clustering techniques.

6.2.2.1 Agglomerative Clustering

Agglomerative Clustering, also known as Hierarchical Clustering, is a clustering algorithm designed to create nested clusters by merging or splitting them successively. @@In the end, these groups of clusters form a are represented in a tree structure, in which the root represents the group of all samples and leaves have only one or a few samples.

The implementation is based on a hierarchical bottom-up approach: each observation starts in its own cluster, and clusters are successively merged together. The linkage criteria can be selected as one of the following: Ward, Maximum linkage, average linkage and single linkage. The criterion determines the metric used for the merge strategy [7].

6.2.2.2 DBSCAN

Another clustering technique adopted is the Density-Based Spatial Clustering of Applications with Noise (DBSCAN). It brings that concept of density and searches for significant samples which are characterized by a high density and starts expanding a cluster from them [8].

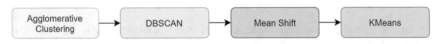

Fig. 6.2 Flow diagram representing the sequence of clustering technique for geographic prediction of the COVID-19

6.2.2.3 Mean Shift

The third clustering method is Mean Shift. It is a centroid-based algorithm, which works by searching for more candidates for centroids to be the mean of the points within a given region and create the clusters. If the selected candidates have close points which can be considered duplicates or similar, then they are excluded to form the final set of centroids [2].

6.2.2.4 K-Means

Finally, the K-Means technique is based on analyzing the variance of some group of samples and clustering similar ones. It minimizes a criterion known as within-cluster sum-of-squares. This algorithm requires the number of clusters to be specified. It scales well to a large number of samples and has been used across a large range of application areas in many different fields [1].

The k-means algorithm divides a set of N samples of X into K disjoint clusters, each described by the mean of the samples in the cluster. The means are commonly called the cluster centroids; note that they are not necessarily, in general, points from X, although they live in the same space.

6.2.2.5 Kalman Filter Prediction

After applying the clustering techniques, the next step is, to consider a two-dimensional Kalman Filter as a predictor and estimate the next set of coordinates/points for the COVID-19 infection [3, 6, 11]. A detailed approach about the Kalman Filter and its characteristics is presented in Chap. 4.

6.3 Results and Discussion

After understanding the design of the proposed methodology of clustering geographic data using different techniques for an effective prediction based on a two-dimensional Kalman filter, this section presents a sequence of results and their corresponding discussion. It is important to notice that the period of the analysis is relevant to be specified so the prediction may be classified as consistent or not.

6.3.1 General Example

The first analysis is a brief general introduction to the visualization system, so the reader can comprehend the proposed approach. Since the number of cases was

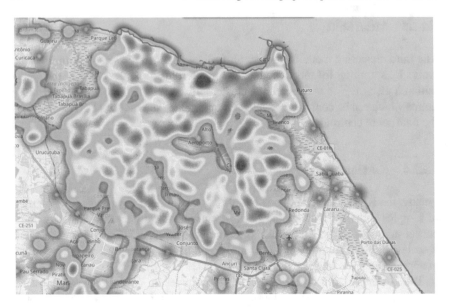

Fig. 6.3 Example of one heatmap of the accumulated number of confirmed cases of COVID-19 in the City of Fortaleza, Brazil, until the end of April 2020

increasing exponentially during the first weeks of the outbreak in Fortaleza, the heatmap plot was selected for data visualization among several other options since this was the best one to allow a clear interpretation of the number of cases without overpopulating the city map.

In Fig. 6.3, it is presented the heatmap of the accumulated number of confirmed cases of COVID-19 in the City of Fortaleza until the end of April 2020. The more red the region, the highest number of cases and the green part represent the lowest number of cases.

After applying the proposed technique of clustering and prediction, in Fig. 6.4 is presented one example of a heatmap and a set of red dots representing the result of geographic predictions of the system, i.e., the locations of possible future occurrences of COVID-19 in the City of Fortaleza, Brazil.

6.3.2   Prediction Case 01—Analyzing the Infection Trend

The first example of prediction will be focused on the most extreme situation, when there were the peaks of new cases per day in the State of Ceara and more specifically in Fortaleza. This occurred from March 15 to April 15, 2020. At that moment, the demand for efficient predictions was high and the support for decision-making was extremely necessary, since the Government needed to allocate resources, establish new rules for controlling physical circulation increasing quarantine measures.

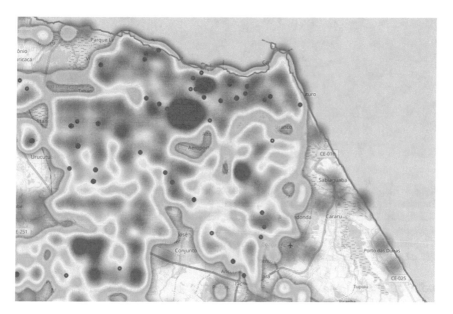

Fig. 6.4 Heatmap and the predicted number of cases, with 1, 500 new possible future occurrences of COVID-19 in the City of Fortaleza, Brazil, until the end of April 2020

@@And, since the beginning of the pandemic, the State Government always based any new decision on scientific data analysis.

In Fig. 6.5 is presented the map of Fortaleza and three levels of zoom. The first one, map (a), is a general overview of the region and the heatmap cannot provide a clear view of the city since it is merging several regions. A group of red dots represents the prediction results. In map (b), it is more feasible to see different clusters of COVID-19 infections over specific neighborhoods. Finally, in map (c), subregions of infection clusters can be found, nevertheless, it demands navigating through the map and zoom out to have a greater overview of the major parts of the city.

About the predictions provided by the system represented by the red dots in the figure, it is important to notice that initially, the cases in Fortaleza were very focused on the northeast region of the city. The challenge at the moment of the prediction was to identify possible paths through which there will have a concentration of new cases per day, according to the clustering of the number of new confirmed cases. The prediction points indicate a trend in the central and northwestern part of the city and this trend was positively confirmed during the following weeks. This real case analysis allowed the Government to prepare the healthcare units of that region for a higher demand in ambulatory support for the population and infrastructure for attending a large demand for COVID-19 diagnostic testing.

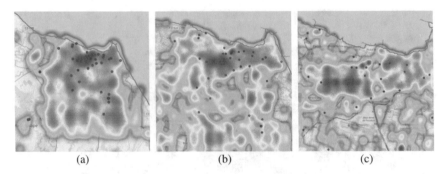

(a)   (b)   (c)

Fig. 6.5 Example of the first case study of the prediction of new COVID-19 cases in the City of Fortaleza, Brazil considering the most significant time of the peak of new cases per day, from March 15 and April 15, 2020. The red dots identify the predicted coordinates of latitude and longitude of possible new cases. **a** General overview of the City with the predictions. **b** More detailed view of the neighborhoods and regions with already with a significant spread in different regions of the City. **c** Deeper vision of one specific region with a set of neighborhoods identified

6.3.3   Prediction Case 02—Importance of the Prediction After the Peak

The second case to be analyzed in this section will focus on the support of the geographic prediction even after the peak of contamination in Fortaleza. The period of this analysis will consider the period between July 1 and 31, 2020. As explained before, since Fortaleza was hit by a large peak of new confirmed cases during the four months before this analysis (March–June), the State and the Municipal Government established several contention measures to reduce the virus infection rates, such as quarantine for the vast majority of the population, closure of public places such as shopping malls, restaurants, schools, and gyms; significant reduction of public transportation and creation of health barriers to control circulation. As a result of these strong measures, during the second half of July, the infection rate started to decrease and the number of new cases significantly reduced.

In Fig. 6.5 is presented the map of Fortaleza and three levels of zoom. The first one, map (a), is a general overview of the region and the heatmap cannot provide a clear view of the city since it is merging several regions. A group of green dots represents the prediction results. Please notice that the color was changed to green and previously the prediction points were red, but since for this analysis the number of cases increased so high and the contamination spread through the whole city, the heatmap plot was almost completely painted in red. In map (b) is allowed the view of specific neighborhoods. Finally, in map (c), subregions can be visualized, nevertheless it demands navigating through the map and zoom out to have a greater overview of other parts of the city.

Analyzing prediction data even after the infection rate has been stabilized is valid for public healthcare administrators, since the possibility of new waves are always under consideration and also the appearance of new regions of concentration of new

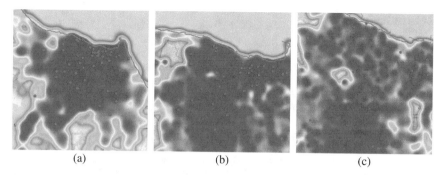

(a) (b) (c)

Fig. 6.6 Second case study of the prediction of new COVID-19 cases in the City of Fortaleza, Brazil considering the period after the peak of new cases per day, between July 1 and 31, 2020. The green dots identify the predicted coordinates of latitude and longitude of possible new cases. **a** General overview of the City with the predictions. **b** More detailed view of the neighborhoods and regions with already with a significant spread in different regions of the City. **c** Deeper vision of one specific region with a set of neighborhoods identified

cases can be identified in advance. For the results presented, still, the central region of the city and the southwest region are identified as critical and that should be actively monitored through healthcare units (Fig. 6.6).

6.4 Conclusion

This chapter presented the case study of a geographic predictor developed to support the decision-making process of the Secretariat of Health of the Government of the State of Ceara, Brazil, during the COVID-19 outbreak.

Two examples were presented in the previous sections. The first one was during the most critical period of the pandemic spread in Fortaleza and the proposed solution could predict correctly where the new cases of the COVID-19 were going to occur, the northwest region of the city. The second example was after a group of severe measures to contain the spread of the virus and the results can be considered as triggers to predict or avoid the appearance of new waves of infection.

This chapter is the conclusion of this book. In the previous chapters, several prediction approaches were considered to analyze and predict the data of the COVID-19 pandemic. The techniques considered three compartmental models (SIR, SEIR, and SEIR with Intervention); linear regressions (ARIMA) model; a quadratic predictor (Kalman Filter) and through more complex models such as LSTM artificial neural networks and auto-selection of machine learning techniques (AutoML).

Although several other prediction methods are being considered in the scientific literature, the methodology and techniques presented in this book intended to provide enough technical information and a critical analysis on the application of each predictor to support the reader in deciding the most suitable and adequate technique.

References

1. A. Ahmad, L. Dey, A k-mean clustering algorithm for mixed numeric and categorical data. Data Knowl. Eng. **63**(2), 503–527 (2007). issn: 0169023X. https://doi.org/10.1016/j.datak.2007.03. 016
2. S. Anand et al., Semi-supervised kernel mean shift clustering. IEEE Trans. Pattern Anal. Mach. Intell. **36**(6), 1201–1215 (2014). issn: 01628828. https://doi.org/10.1109/TPAMI.2013.190
3. M. Gatto et al., Spread and dynamics of the COVID-19 epidemic in Italy: Effects of emergency containment measures. Proc. Natl. Acad. Sci. U. S. A. **117**(19), 10484–10491 (2020). issn: 10916490. https://doi.org/10.1073/pnas.2004978117
4. M.F. Goodchild, P. Fu, P. Rich, Sharing geographic information: an assessment of the geospatial one-stop. Ann. Assoc. Am. Geogr. **97**(2), 250–266 (2007). issn: 00045608. https://doi.org/10. 1111/j.1467-8306.2007.00534.x
5. M.F. Goodchild, J. Alan Glennon, Crowdsourcing geographic information for disaster response: a research frontier. Int. J. Digit. Earth **3**(3), 231–241 (2010). issn: 1753- 8947. https://doi.org/10.1080/17538941003759255. http://www.tandfonline.com/doi/abs/10.1080/ 17538941003759255
6. A. Monfort, J.-P. Renne, G. Roussellet, A Quadratic Kalman Filter *. Tech. rep (2013)
7. D. Müllner, Modern hierarchical, agglomerative clustering algorithms (2011). arXiv:1109.2378. http://arxiv.org/abs/1109.2378
8. E. Schubert et al., DBSCAN revisited, revisited: why and how you should (still) use DBSCAN. ACM Trans. Database Syst. **42**(3), 1–21 (2017). issn: 15574644. https://doi.org/10.1145/ 3068335. https://dl.acm.org/doi/10.1145/3068335
9. S. Zhao et al., The association between domestic train transportation and novel coronavirus (2019-nCoV) outbreak in China from 2019 to 2020: a data-driven correlational report (2020). https://doi.org/10.1016/j.tmaid.2020.101568
10. C. Zhou et al., COVID-19: Challenges to GIS with big data. Geogr. Sustain. **1**(1), 77–87 (2020). issn: 26666839. https://doi.org/10.1016/j.geosus.2020.03.005
11. Q. Zhuang, Application of Kalman filtering in dynamic prediction for corporate financial distress, in *Kalman Filters - Theory for Advanced Applications*. InTech (2018). https://doi.org/ 10.5772/intechopen.71616. http://dx.doi.org/10.5772/intechopen.71616